Alexa®

for
dummies®
A Wiley Brand

Alexa®

2nd Edition

by Paul McFedries

A Wiley Brand

Alexa® For Dummies®, 2nd Edition

Published by **John Wiley & Sons, Inc.**, 111 River Street, Hoboken, NJ 07030-5774, www.wiley.com

Contents at a Glance

Table of Contents

Introduction

O kay, I know just what you're thinking: Why, oh, why would someone write a book about Alexa, Amazon's famous and wildly popular virtual assistant? After all, don't you just plug in an Echo or any of the umpteen other devices that come with Alexa built in, say, "Alexa," and then ask a question or give a command? What else does a person need to know, for pity's sake? I'm surprised Alexa requires even *three* pages of explanation, so how can this book have hundreds?

First, wow, you ask a lot of questions. Second, well, you'd be surprised. Sure, Alexa's basics are readily understood. That's part of the beauty of Alexa and the devices that run it: The initial learning curve is pretty much a horizontal line.

Ah, but Alexa is *way* deeper and more powerful than those out-of-the-box basics would indicate. You have a serious piece of software at your beck and call. (Well, your call, anyway; Alexa does not yet respond to becks.) I'm talking high-end artificial intelligence (AI), world-class voice recognition, and tons of third-party extensions. All that adds up to an amazingly sophisticated app that contains hidden depths just waiting for you to find them.

About This Book

Alexa For Dummies, 2nd Edition not only takes you on a complete tour of Alexa's capabilities, tools, settings, and skills but is also fully updated to cover all the latest Alexa devices and features. You'll even see that Alexa has a hidden whimsical side that it's just dying to show. In the end, you'll discover everything you need to know to get the most out of your Alexa investment. Best of all, you'll have a ton of fun as you go along.

This book boasts 15 chapters, but just because they appear sequentially, that doesn't mean you have to read them that way. Use the table of contents or index to find the information you need — and dip into and out of the book when you have a question about Alexa.

If your time is limited, you can also ignore anything marked by the Technical Stuff icon or the information in sidebars (the gray-shaded boxes). Yes, these tidbits are fascinating, but they aren't critical to the subject at hand, so you won't miss anything critical by skipping them.

Within this book, you may note that some web addresses break across two lines of text. If you're reading this book in print and want to visit one of these web pages, just key in the web address exactly as it's noted in the text, as though the line break doesn't exist. If you're reading this as an ebook, you've got it easy — just click the web address to be taken directly to the web page.

Finally, throughout this book, when you see an italicized word in a command to Alexa, that's a placeholder. That is, you replace the italicized word with something else specific to your command. For example, the placeholder *title* means you replace *title* with the actual title of something (such as a song or an album).

Foolish Assumptions

This book is for people who are new (or relatively new) to using the Alexa voice-activated virtual assistant. Therefore, I do *not* assume that you're an Alexa expert, an Alexa maven, or an Alexa aficionado. However, I do assume the following:

» You know how to plug in and connect devices.

» You have a running Wi-Fi network with an Internet connection.

» You know the password for your Wi-Fi network.

» You have either an iOS or an Android mobile device (that is, a smartphone or tablet).

» You know how to install and operate apps on your mobile device.

Icons Used in This Book

Like other books in the *For Dummies* series, this book uses icons, or little pictures in the margin, to flag things that don't quite fit into the flow of the chapter discussion. Here are the icons that I use:

REMEMBER

This icon marks text that contains something useful or important enough that you'd do well to store the text somewhere safe in your memory for later recall.

TECHNICAL STUFF

This icon marks text that contains some for-nerds-only technical details or explanations that you're free to skip.

TIP

This icon marks shortcuts or easier ways to do things, which I hope will make your life — or, at least, the Alexa portion of your life — more efficient.

WARNING

This icon marks text that contains a friendly but unusually insistent reminder to avoid doing something. You have been warned.

Beyond the Book

In addition to what you're reading right now, this product also comes with a free access-anywhere cheat sheet that includes ten must-do setup tasks, 50 of Alexa's most useful voice commands, and 50 of Alexa's most fun voice commands. To get this cheat sheet, go to www.dummies.com and type **Alexa For Dummies Cheat Sheet** in the Search box.

Where to Go from Here

If you've had Alexa for a while and you're familiar with the basics, you can probably get away with skipping the first three chapters and then diving in to any part of the book that tickles your curiosity bone. (Or, why not take advantage of your handy personal assistant? Just say, "Alexa, give me a number between 4 and 15.") The chapters all present their Alexa know-how in readily digestible, bite-size tidbits, so feel free to graze your way through the book.

However, if you and Alexa haven't met yet — particularly if you're not sure what Alexa even *does* — this book has you covered. To get your relationship with Alexa off to a fine start, I highly recommend reading the first three chapters to get some of the basics down cold. From there, you can branch out anywhere you like, safe in the knowledge that you have some survival skills to fall back on!

1
Getting Started with Alexa

Find out what Alexa is, what Alexa can do, and what hardware you need to use Alexa.

Welcome Alexa into your home by learning where to put your Alexa device, getting your device on your network, and customizing Alexa.

Discover some crucial basics about your Echo device and Alexa itself, as well as some important Alexa commands.

Chapter **1**

Getting to Know Alexa

An old proverb tells us that "Well begun is half done." That is, if you begin a project in the best way you can, you'll have made such a good start that it'll seem like you're already halfway to your goal. In this chapter, you begin your relationship with Alexa well by learning some useful, perhaps even interesting, background about Alexa, including an answer to what might be the most important question of all: Just what *is* Alexa, anyway? (Or should that be just *who* is Alexa?)

To get your Alexa education off to a solid start, this chapter takes you on an exploratory tour of the Alexa landscape. This is big-picture stuff where you learn not only what Alexa is but also where you can get Alexa and how to figure out which Alexa-friendly device you need. After taking you through these *what, where,* and *how* fundamentals, you also investigate what is likely the *second* most important question: *Why* would people even need Alexa in their lives?

What Is This Alexa That Everyone's Talking About (or, Really, To)?

Okay, let me get right to it: Amazon Alexa is a *voice service,* a cloud-based software program that acts as a voice-controlled virtual personal assistant. In a nutshell, you use your voice to ask Alexa a question or give Alexa a command, and it dutifully answers you (assuming an answer exists) or carries out your request (assuming your request is possible). The key here is that Alexa responds to *voice* commands.

In the movie *Star Trek IV: Voyage Home,* the crew of the Starship *Enterprise* travel back in time 300 years to 1986. In a memorable scene, Scotty, the ship's chief engineer, goes up to a mid-'80s-era PC and says, "Computer!" When the machine doesn't respond, he says, "Computer!" once again. He's then handed a mouse and, thinking it's a microphone, says, "Hello, computer!" Apparently, in the year 2286, interacting with a computer using anything but voice commands is unthinkable.

We're a long way from the voice-only future envisioned in *Star Trek* (and countless other sci-fi stories; remember voice-controlled HAL in *2001: A Space Odyssey?*). However, as we sit here near the end of the second decade of the 21st century, you can feel the computer-interaction landscape starting to shift. After some 40 years of folks sitting in front of their PCs, typing away in near-total silence (with only the occasional wail of exasperation or groan of impatience to break the quiet), users are starting to find their voices.

True, operating systems such as Windows and macOS have had voice-control tools for many years, but they were obscure and unreliable and used by only a handful of people. Voice control's bid for the mainstream didn't get serious until Apple purchased the Siri speech-recognition app in 2010 and released it with iOS 5 a year later. Suddenly, it became cool to interact with a computer (at least one in the shape of a smartphone) by using voice commands.

Since then, numerous voice-control tools have been released, including Google's Assistant, Microsoft's Cortana, and various voice-command features found in modern cars. But it was the release of the full version of Amazon's Alexa in 2015 that really got the voice ball rolling. Amazon doesn't share sales figures, but most industry know-it-alls agree that at least two hundred million Alexa devices have been sold.

Meet your new assistant

Why is Alexa so popular? There are lots of reasons, but the one that matters is that Alexa is (or tries hard to be) a personal assistant. Older voice-command tools were geared toward using a computer: running programs, pulling down menus, accessing settings, and so on. Alexa doesn't do any of that. Instead, it's focused on doing things for you in your real life, including (but by no means limited to) the following:

- » Playing music, podcasts, or audiobooks
- » Setting timers and alarms
- » Telling you the latest news, weather, or traffic
- » Creating to-do lists and shopping lists
- » Buying something from Amazon
- » Controlling home-automation devices such as lights and thermostats
- » Telling jokes

That last one may be a bit surprising, but perhaps the second-most important reason behind's Alexa's success is that it comes with whimsy as a feature. Alexa, as I hope to show in this book, is both useful *and* fun.

REMEMBER

Older and lesser voice-controlled systems recognize only a limited set of commands that have to be enunciated precisely, so using such systems feels stilted and slow. Alexa, by contrast, is an example of a new breed of voice-aware systems that use *conversational artificial intelligence.* That term sounds pretty geeky, but it simply means that Alexa isn't meant to be controlled so much as *interacted* with. With Alexa, you ask your questions and give your commands using natural language and your normal voice. Does it work perfectly every time? Nope, we're not in *Star Trek* territory just yet, but I think you'll be pleasantly surprised at just how well it *does* work.

Alexa's components

Throughout this book I talk about "Alexa" as though it's a single object, but Alexa is actually a large collection of objects that, together, create the full, seamless Alexa experience. I talk about many of these objects — particularly the Alexa app — throughout this book. From the behind-the-scenes perspective, however, all you need to know for now is that Alexa consists of the following four components:

- » **Name recognition:** When you interact with Alexa, it seems as though the device understands what you say, but the only speech your Alexa device recognizes is the word *Alexa,* which Amazon calls the *wake word.* That is, it's the word that lets Alexa know it should wake up and start listening for an incoming command or question. (In case you're wondering: Yep, you can change the wake word to something else. I show you how to do that in Chapter 15.)

- » **Speech recording:** Your Alexa device has one or more built-in microphones that capture the questions, commands, requests, and other utterances that you direct to the device. A simple computer inside records what you say and then sends the recording over the Internet to the Alexa Voice Service (discussed next). This part of Alexa is sometimes called the *voice user interface* (VUI).

» **Alexa Voice Service (AVS):** Here's where the real Alexa magic happens. This part of Alexa resides in Amazon's cloud. AVS takes the recording that contains your voice command and uses some fancy-schmancy speech recognition to tease out the words you spoke. AVS then uses natural-language processing to analyze the meaning of your command, from which it produces a result.

» **Speech synthesis:** This component takes the results provided by AVS and renders them as speech, which it stores in an audio file. That file is returned and played through the Alexa device's built-in or connected speakers.

TECHNICAL STUFF

WHAT'S ALL THIS ABOUT A "CLOUD"?

I've mentioned the term *cloud* a couple of times now, so let me take a few minutes of your precious time to explain what I'm talking about. In many network diagrams, the designer is most interested in the devices that connect to the network, not the network itself. After all, the details of what happens inside the network to shunt signals from source to destination are often complex and convoluted, so all those minutiae would only detract from the network diagram's larger message of showing which devices can connect to the network, how they connect, and their network entry and exit points.

When designers of a network flowchart want to show the network but not any of its details, they almost always abstract the network by displaying it as a cloud symbol. (It is, if you will, the "yadda yadda yadda" of network diagrams.) At first the cloud symbol represented the workings of a single network, but in recent years it has come to represent the Internet (the network of networks).

So far, so good. Earlier in this millennium, some folks had the bright idea that instead of storing files on local computers, they could be stored on a server connected to the Internet, which meant that anyone with the right credentials could access the files from anywhere in the world. Eventually, folks started storing programs on Internet servers, too, and started telling anyone who'd listen that these files and applications resided "in the cloud" (meaning on a server — or, more typically, a large collection of servers that reside in a special building called a *data center* — accessible via the Internet).

One such application is Alexa Voice Service, which resides inside Amazon's cloud service called Amazon Web Services (AWS). So, that's why I say that Alexa is a "cloud-based voice service." That's also why you need an Internet connection to use Alexa: It requires that connection to access its cloud component.

How Alexa works

Given the various Alexa components that I outline in the preceding section, here's the general procedure that happens when you interact with Alexa to get something done:

1. **You say "Alexa."**

   Your Alexa device wakes up, as indicated (in most cases) by the device's light ring turning blue. (See Chapter 3 for more about the light ring.) The device is now hanging on your every word.

2. **You state your business: a question, a command, or whatever.**

   The Alexa device records what you say. When you're done, the device uses your Internet-connected Wi-Fi network to send the recording to AVS in Amazon's cloud.

   It may seem as though Alexa lives inside whatever device you have, but Alexa is an Internet-based service. If you don't have Internet access, you don't have Alexa access either.

3. **AVS uses its speech-recognition component to turn the recorded words into data that can be analyzed.**

4. **AVS uses its natural-language processing component to analyze the words in your command and then try to figure out exactly what you asked Alexa to do.**

   AVS doesn't analyze every single word you say. Instead, it's mostly looking for the telltale *keywords* that indicate what you've asked Alexa to provide. For example, if you said "What's the weather forecast for tomorrow?," all AVS needs are the words *weather, forecast,* and *tomorrow* to deliver the correct info.

5. **If AVS can't fulfill your request directly, it passes the request along to a third-party service (such as AccuWeather or Wikipedia), and then gathers the response.**

6. **AVS returns the response via the Internet to your Alexa device.**

   What AVS returns to your Alexa device depends on the result. If the result is just information for you, AVS passes along the text and also converts the text to speech and stores the speech in an audio file that your Alexa device can play. If the result is a command (for example, to play a particular song), AVS passes that command back to the Alexa device.

7. **The Alexa device either uses its built-in or connected speaker to broadcast the result of your request or carries out your command and lets you know what it did.**

   If your Alexa device has a screen, you also see the result on the screen.

Where Do You Get Alexa?

In most cases, Alexa is closely associated with hardware devices, and how close that association is depends on the device. There are three types of devices to consider:

» **Devices that have Alexa built in:** As you might have expected, Amazon offers a huge range of products that have Alexa inside, including the Echo, Echo Dot, Echo Plus, Echo Spot, Echo Show, Fire tablet, and Fire TV. In addition, a massive ecosystem of third-party devices are Alexa-enabled, including select Windows PCs, sound systems, TVs, tablets, appliances, GPS units, and even car models.

» **Devices that Alexa can control directly:** Many devices are Alexa-friendly, meaning that Alexa can connect to and control those devices directly using either a Bluetooth or Wi-Fi connection. For example, you can use Amazon's Echo Show to connect to and control a smart-home device such as a thermostat. Similarly, you can use any Amazon Echo device to operate the AmazonBasics microwave (yes, that's right: a microwave oven that you can control with your voice!).

» **Devices that Alexa can control indirectly:** Although as I write this some 100,000 devices fit into the previous two categories, that still leaves a huge number of Alexa-ignorant devices. However, in some cases you can still have limited control over even these devices. For example, you can connect any Amazon Echo device to an Amazon smart plug, which you can turn on and off using Alexa. So, when you plug a non-Alexa device (such as a lamp) into the smart plug, you can use Alexa to turn that device on and off.

Note that, at the start of this section, I said, "in most cases." What are the exceptions? As I talk about in Chapter 2, Amazon offers the Alexa app, which is a program that you install on your smartphone or tablet. You can use the app to connect with and manage smart-home devices, but if you just want the standard Alexa experience — that is, using voice commands to ask questions and make requests — you can do all that directly from the app; no external hardware (such as the devices I ramble on about in the next section) is required.

Figuring Out Which Alexa Device You Need

In the preceding section, I mention that, with the exception of using Alexa on your phone or tablet, you can't do the Alexa thing until you get a device that's Alexa-enabled. That sounds straightforward enough, but that illusion of simplicity is shattered when you see the sheer number of available devices — and I'm not even talking about all the third-party Alexa devices. Amazon alone offers well over a dozen different Alexa-enabled devices just in its Echo brand of smart speakers. How are you supposed to know which one to get?

Checking out Amazon's Echo devices

To help you make the right Alexa decision, this section offers a quick look at what's available from Amazon's Echo brand.

Echo

Echo (the fourth generation is shown in Figure 1-1) is your garden-variety Echo smart speaker designed for larger rooms because it comes with three speakers: a 3-inch woofer and two 0.8-inch tweeters. It's fairly big — 5.7 inches in diameter and 5.2 inches high — so you may need to clear a spot for it. The fourth generation of the Echo comes with a built-in smart-home hub.

FIGURE 1-1:
Amazon's Echo smart speaker.

Photograph courtesy of Amazon

Echo Dot

The Echo Dot is a smart speaker designed for smaller rooms because it comes with a single speaker. It's quite a bit teensier overall than the Echo (about 3.9 inches in diameter and about 3.5 inches high). It's also half the price of the Echo, which is likely why it's Amazon's bestselling Alexa device. Some models of the fourth generation Echo Dot also come with a built-in clock, as shown in Figure 1-2.

Photograph courtesy of Amazon

FIGURE 1-2:
Amazon's Echo
Dot smart
speaker.

Echo Studio

Echo Studio (shown in Figure 1-3) is the smart speaker designed for audiophiles thanks to its support of several high-end audio formats (including Dolby Atmos) and five — yep, you read that right: *five* — speakers: a 5.3-inch woofer, a 1-inch tweeter, and three 2-inch midrange speakers. It's the biggest Echo at 6.9 inches in diameter and 8.1 inches high, so it won't go unnoticed. Echo Studio also comes with a built-in smart-home hub.

Echo Flex

The Echo Flex (shown in Figure 1-4) is a smart speaker that plugs directly into an electrical outlet. The Echo Flex comes with a single speaker (so it's not suitable for music playback) and is 2.6 inches wide and 2.8 inches high. Yep, that's tiny, but Echo Flex has a USB port that you can use to charge your USB devices or connect accessories such as a night light or a motion sensor.

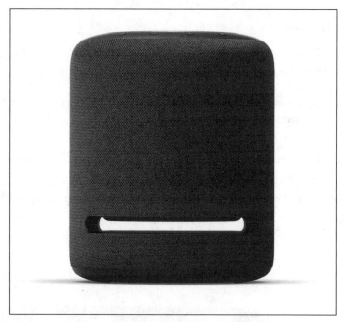

FIGURE 1-3:
Amazon's Echo
Studio smart
speaker.

Photograph courtesy of Amazon

FIGURE 1-4:
Amazon's Echo
Flex plug-in smart
speaker.

Photograph courtesy of Amazon

Echo Show

The Echo Show is a smart speaker that comes with a screen. Three models are available:

» **Echo Show 5:** Comes with a 5.5-inch (measured on the diagonal) screen that supports 960x480 resolution; a 1-megapixel (MP) camera; and a single 1.7-inch speaker. See Figure 1-5.

» **Echo Show 8:** Has an 8-inch screen and 1280x800 resolution; a 1MP camera; and two 2-inch speakers.

» **Echo Show 10:** Offers a 10.1-inch and 1280x800 resolution; a 13MP camera; and three speakers (a 3-inch woofer and two 1-inch tweeters); and a built-in smart-home hub.

FIGURE 1-5:
Amazon's Echo Show 5 smart speaker.

Photograph courtesy of Amazon

Echo Auto

The Echo Auto (shown in Figure 1–6) is designed to hang out with you in your car. It connects to the Internet via your smartphone. The Echo Auto doesn't have a built-in speaker, but you can use it to connect to your car's speakers via Bluetooth or by plugging in an audio cable (assuming your car supports either of these options). The Echo Auto is 3.3 inches wide and 1.9 inches deep.

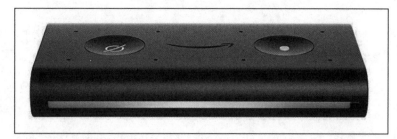

FIGURE 1-6:
Amazon's Echo
Auto device.

Echo Frames

Echo Frames (see Figure 1-7) is a pair of smart glasses that somehow manages to shoehorn two microphones and four teensy speakers (called *microspeakers*) into what look like a regular pair of reading glasses. (You supply the lenses, by the way.) Echo Frames uses the Alexa app on your iOS, iPadOS, or Android mobile device to connect to the Internet to process your voice commands and to return the results of those requests.

FIGURE 1-7:
Amazon's Echo
Frames smart
glasses.

Choosing an Alexa device

Okay, so which one should you choose? The answer depends on your needs and budget, so here are some questions to help you decide:

>> **Do you just want to ask questions and make requests?** If you have a smartphone or tablet that's Alexa-friendly (I talk about what that means in Chapter 2), don't bother getting any Alexa device. Just use the Alexa app instead.

>> **Are you on a tight budget?** Go for the Echo Flex, which is by far the cheapest of the Echo devices.

>> **Do you want decent sound without breaking the bank?** Get the Echo, which comes with both a woofer and a tweeter and is priced midway between the Echo Dot and the Echo Studio.

>> **Do you want the best sound possible?** Buy the Echo Studio, with its five speakers. If your budget can handle it, you might want to add an Echo Sub, which is a powerful subwoofer (see Chapter 12).

>> **Do you want Alexa available in multiple rooms?** Most people who go this route have an Echo or Echo Studio in a main area (such as the kitchen or living room) and multiple Echo Dots or Echo Flexes: one in each room where you want to use Alexa.

>> **Do you want to use Alexa to manage smart-home devices?** In this case, you need the Echo, Echo Studio, or Echo Show 10, each of which has a smart-home hub built in.

>> **Do you want to use Alexa while you're rambling around town?** You can use the Alexa app on your iOS or Android smartphone, but if you also prefer to use Alexa hands-free, you'll want a pair of Echo Frames.

>> **Do you want to make video calls?** You need an Echo with both a screen and a camera, so that means one of the Echo Show devices.

Learning What Alexa Can Do

Some people think an Alexa-enabled device is nothing but a glorified clock-radio. Sure, it's true that some versions of the Echo Dot and all versions of the Echo Show can display the time (and all Alexa devices can tell you the time) and you can use Alexa to play a radio station. However, Alexa can do more — *way* more. Let me show you what I mean.

Playing media

You can connect your Alexa device to your favorite music service — such as Amazon Music or Spotify — which gives you voice-controlled access to millions of songs. You can ask Alexa to play a particular song, artist, playlist, or genre. If you have multiple smart speakers in your home, Alexa's multiroom music feature enables you to play the same music in each room that has a speaker. Alexa can also play radio stations, podcasts, and audiobooks; recite the text of a Kindle book; and, if your device has a screen, play shows from Amazon Prime Video. See Chapter 4 to get the scoop on Alexa's media playback features and Chapter 12 to learn about multiroom music.

Communicating

One of Alexa's more surprising features is that you can use it to communicate with other people, even if they don't have an Alexa device. You can send text messages and place voice calls, but the fun begins when you and the other person have Alexa devices with screens because then you can make video calls to that person. If you have multiple Alexa devices at home, you can also use them as a two-way intercom system or as a one-way intercom for making announcements. Some restrictions apply, so be sure to check out Chapter 5 to learn more.

Shopping

It wouldn't be an Amazon product if shopping weren't involved, so of course you can use Alexa to place orders through your Amazon account, check the latest Amazon deals, and track your shipments. I cover shopping details in Chapter 6.

Getting help around the home

We're all busy, so who couldn't use an extra hand around the house now and then? Alexa has no hands, alas, but it can help big-time by making it easy to create a to-do list or shopping list; manage your Google, iCloud, or Office calendar; and set reminders for upcoming tasks. Alexa can also set an alarm and run a timer. I talk about all these features and more in Chapter 7.

Getting news and information

Alexa is always up on the latest news, so all you have to do is ask. You can even customize the news you hear by configuring Alexa's Flash Briefing feature. Alexa can also tell you the current weather and the latest forecast, give you traffic updates, let you know the scores and schedules of your favorite sports teams, and tell you movie showtimes at nearby theaters. Alexa is a veritable fount of information, as I describe in more detail in Chapter 6.

Answering your questions

If there were such a thing as a Swiss Army knife for information, Alexa would be it. I've already mentioned that it's a news anchor, a weather forecaster, and a sportscaster, but Alexa can also be a calculator, a speller, dictionary, an encyclopedia, and a search engine. Ask a question, and the odds are in your favor that Alexa can find the answer. I talk about Alexa's Q&A prowess in Chapter 8.

Accessing skills

Alexa comes with tons of built-in features, but Amazon has set up a system that lets third parties add new features to Alexa. These features are called *skills*, and thousands of them are available in Amazon's Alexa Skills Store. There are skills for ordering a pizza or an Uber, playing games or trivia, tracking your fitness or your investments, and so much more. You can even build your own skills without programming! I take you to skills school in Chapter 9.

Automating your home

Alexa is smart-home savvy, so it gives you voice control over many different home-automation products, including lights, thermostats, baby monitors, security cameras, and door locks. I tell you how to set up and manage your smart home in Chapters 11 and 12.

Having fun

With its easy access to news and weather, its to-do lists and reminders, and its massive catalog of life-hacking skills, Alexa may seem as though it's all business, but it has a fun side as well. Alexa can tell jokes, read limericks, sing songs, tell stories, and play games. Alexa even comes with a large trove of so-called Easter eggs that bring surprise and whimsy to your Alexa conversations. Loosen your tie or your let down your hair, and then head for Chapter 13 to learn more.

> » **Connecting Alexa to your Wi-Fi network**
>
> » **Customizing your Alexa device**
>
> » **Connecting external speakers or headphones**

Chapter **2**

Setting Up Alexa and Your Devices

The basic premise of Alexa couldn't be simpler: You ask a question, Alexa provides the answer; you give a command, Alexa carries it out. However, this surface simplicity lies on top of a mind–bogglingly complex system that involves hardware, software, networking, artificial intelligence, and the cloud. (I cover the details of this system in Chapter 1.) Fortunately, that underlying complexity isn't something you have to deal with, but before you can use Alexa, you need to connect a few things so that Alexa has access to all that good stuff beneath the surface.

In this chapter, you delve into the various chores — both required tasks and a few optional steps— that turn your Alexa device from a useless hunk of plastic into a useful member of the family.

Installing the Alexa App

Here are a few questions about Alexa that may or may not have occurred to you:

> » Alexa requires an Internet connection, but how does Alexa know how to connect to your Wi-Fi network?

>> Alexa can tell you the local news, weather, and traffic, but how does Alexa know where you live?

>> Alexa can work with other devices such as lights and thermostats, but how does Alexa know when such devices are present?

>> Alexa can play music, but how does Alexa know whether you use Amazon Music or Spotify (or Pandora or whatever)?

>> You can't wait to use Alexa to buy stuff, but how does Alexa know about your Amazon account?

Excellent questions all around, and they all have the same answer: the Alexa app, which is a program you download to your smartphone or tablet. With the Alexa app, you can connect your Alexa devices to your Wi-Fi network, provide Alexa with your location, connect to smart-home devices, let Alexa know which music service you use, provide the details of your Amazon account, and much more. You *use* Alexa by conversing with an Alexa-friendly device, but you *configure* Alexa by using the app.

Okay, so what do you need to get the Alexa app? Either of the following:

>> **A smartphone or tablet that meets one of these qualifications:**

- An iPhone running iOS 11 or later

- An iPad running iPadOS 11 or later

- An Android phone or tablet running Android 6 or later

- An Amazon Fire tablet running Fire OS 5.3.3 or later

If you have one of these devices, go to your device's app store, search for the Alexa app, and install it.

If you have one of Amazon's Fire tablets and that tablet is capable of running Alexa, the Alexa app will be installed automatically on the tablet.

REMEMBER

>> **A Windows or Mac web browser:** In this case, point your browser to https://alexa.amazon.com to go to the Amazon Alexa portal page.

Either way, your first app task is to sign in to your Amazon account, as I describe next.

Signing in to your Amazon account

When you first open the Alexa app or surf to the Amazon Alexa portal page, you're prompted to sign in with your Amazon account credentials. Figure 2-1 shows the Sign In page for the Alexa app on an iPhone.

FIGURE 2-1:
The iPhone
version of the
Alexa app's Sign
In page.

REMEMBER

When you open the Alexa app, it might ask for permission to use Bluetooth. That's cool, so go ahead and tap OK or Allow.

Follow these steps to sign in:

1. **In the Email field, type your Amazon account's email address.**

   If you use an Amazon mobile phone account instead of a standard Amazon account, type your mobile phone number into the field instead of your email address.

   What if you don't have an Amazon account? No problem. Tap the Create a New Amazon Account button; use the Create Account screen to type your name, email address, and a password; and then tap Create Your Amazon Account.

2. **In the Amazon Password field, type your Amazon account's password.**

   By default, the Alexa app hides the password by displaying each character as a dot. If you want to make sure you've entered the password correctly, select the Show Password check box.

3. **If you are using the Amazon Alexa portal page and want to bypass signing in on subsequent visits, select the Keep Me Signed In check box.**

4. **Tap the Sign-In button.**

The Alexa app confirms your Amazon credentials and then signs in to your account. The app then asks you to confirm your name or create a new name to use with Alexa.

REMEMBER

If you've enabled two-factor authentication on your Amazon account (as I describe in Chapter 15), the Alexa app will prompt you to enter a code to verify the sign-in. Type the code that was sent to you and then tap the Sign-In button.

5. **Tap I'm *Name,* where *Name* is the name associated with your Amazon account.**

To use a different name, tap I'm Someone Else, enter the new name in the First Name and Last Name fields, and then tap Continue.

If the app shows a screen telling you about new features, say "Whatever, dude" and tap Done.

The app now prompts you for permission to access your contacts (Android, iOS, and iPadOS) and to display notifications on your device (for example, when you receive a text message or an alert such as a reminder; iOS and iPadOS only).

6. **Tap Allow and, when you're asked to confirm, tap Allow (Android) or OK (iOS or iPadOS).**

The Alexa app prompts you to create a voice profile, which enables Alexa to learn your voice and your name.

7. **Tap Continue.**

Alexa asks for permission to use your device's microphone.

8. **Tap Allow (Android) or OK (iOS or iPadOS).**

9. **Follow the instructions provided by Alexa.**

Alexa greets you, asks your name, and then asks you to repeat a bunch of phrases such as "Alexa, what are your deals?" and "Amazon, search for holiday music."

10. **When you're done, tap Next or OK to finish the setup.**

The Alexa app finishes loading and you see the Home screen, similar to the one in Figure 2-2.

FIGURE 2-2:
The Alexa app's
Home screen.

Taking a tour of the Alexa app

In Figure 2-2, you can see a few landmarks of the Alexa app screen. Here's a summary:

» **Alexa:** Tap the icon labeled Tap to Talk to Alexa at the top of the Home screen to send questions and requests to Alexa via your smartphone or tablet microphone. See "Giving Alexa access to your device microphone," next.

» **Home:** Tap this icon to display the Home screen, which offers the Things to Try section (suggestions for getting started with Alexa) and (further down) a series of sections — known as *cards* — that present in reverse chronological order (that is, the newest at the top) your most recent interactions with Alexa: your questions and Alexa's answers, responses to your requests, recently played music and other media, and more.

» **Communicate:** Tap this icon to open the Communication screen. Here, you can set up and work with Alexa's communications features, including starting a new call, sending a text message, and dropping in on someone.

» **Play:** Tap this icon to see what media — such as a song, a podcast, or an audiobook — you've played recently on your Alexa device. You can also access your music services — such as Amazon Music, Spotify, and TuneIn — and browse your Audible and Kindle libraries.

>> **Devices:** Tap this icon to work with both your Alexa devices and Alexa's home-automation features. You can add, configure, and operate devices, manage smart-home skills, create smart-home groups, and more.

>> **More:** Tap this icon to access the main app menu, shown in Figure 2-3. You use the commands on this menu to configure various Alexa features, such as reminders and alarms; add and manage Alexa skills; and change the Alexa app's settings. I cover all these commands in the appropriate sections of the book. (For example, I cover the Alarms & Timers command in Chapter 7 and the Settings command later in this chapter.)

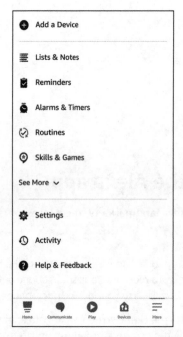

FIGURE 2-3:
Tap the More icon to see this menu of app commands.

Giving Alexa access to your device microphone

One of the nice features of the Alexa app is that you can use it to send voice commands to Alexa via your smartphone or tablet microphone. That's the purpose of the Alexa icon at the top of the Home screen (refer to Figure 2-2). However, before you can send voice requests to Alexa, you need to give the Alexa app permission to use your device's microphone. Here's how it works:

1. **In the Home screen of the Alexa app, tap the Alexa icon (refer to Figure 2-2).**

   You see some text telling you that you need to give the app permission to use the microphone.

2. **Tap Allow.**

   Your device asks you to confirm.

3. **Tap Allow (Android) or OK (iOS or iPadOS).**

   Your device gives the Alexa app permission to access the microphone and then displays some examples of things you can say to Alexa.

   If you also see a prompt asking you to give the Alexa app permission to use your location, go ahead and tap Allow.

4. **Tap Done.**

   You now hear a tone and see the regular Alexa screen, which is waiting for you to say something to Alexa. You'll want to skip that step for now, so either tap the X to close the screen or wait a few seconds for the screen to close by itself.

Enabling the Alexa app's hands-free mode

As you see throughout the rest of this book, one of nice features of most Alexa-friendly devices is their "Look, ma, no hands!" convenience. That is, you say "Alexa," state your request, and it all just works. You don't have to press or hold down anything; you just use your voice.

Unfortunately, you might not have that same convenience with the Alexa app. That is, you might have to first tap Home, then tap the Alexa icon, then state your request. Too many taps! A better way to use the Alexa app is to activate its hands-free mode, which enables you to use voice-only commands to make Alexa do your bidding.

Hands-free mode is usually activated by default. How can you tell? Open the Alexa app's Home tab and check the text below the Alexa icon:

>> **Tap or say "Alexa":** If you see this text, hands-free mode is activated.

>> **Tap to talk to Alexa:** If you see this text, hands-free mode is not activated. Follow these steps to turn on hands-free mode:

   1. *Tap More.*

   2. *Tap Settings.*

3. *Tap Alexa App Settings.*

4. *Tap the Enable Alexa Hands-Free switch to On.*

When hands-free mode is enabled, you can talk directly to Alexa whenever you're using the app. You don't have to be in the Home tab and you don't have to tap the Alexa icon. Nice.

Alexa Feng Shui: Positioning Your Alexa Device

Assuming you've liberated your Alexa device from its packaging, one big question remains: Where do you put it? That's a vexing question for two reasons:

>> Unlike a smartphone, tablet, or laptop, most Alexa devices aren't portable because they need to be plugged in to a power outlet.

>> Unlike a desktop computer, which usually has an obvious location (that would be the top of your desk!), Alexa devices have no such go-to place.

So, again, where do you put it?

REMEMBER

Out of the box, almost all Alexa devices require a power outlet, but third-party gadgets enable devices such as the Amazon Echo to run on batteries. For some examples, search Amazon for *battery base Echo*.

Here are some things to consider:

>> You'll want the whole family to have easy access to the Alexa device, so choose a room where everyone hangs out. This might be the family room or the den, but I'm betting the kitchen will be the best choice.

WARNING

If you decide the kitchen is the best place for your Alexa device, be sure to position it well away from your microwave oven. Microwaves can wreak havoc on the Wi-Fi signals that Alexa depends upon.

>> Make sure the Alexa device is within range of your Wi-Fi network.

>> Make sure the device is close enough that you can give voice commands without having to yell. Depending on the ambient noise in your environment, you usually need to be within 15 to 20 feet of the device.

>> For the best sound quality, keep the device at least 8 inches from any wall.

>> Alexa devices require full-time power, so make sure an outlet is close enough to the device.

With your device positioned where you want it, go ahead and plug it in. After a brief warm-up, the device will let you know that it's ready to be set up using the Alexa app.

Adding Your Alexa Device

With the Alexa app installed and your Alexa device positioned where you want it, you're ready to add the device to the Alexa app so that you can customize and control the device through your smartphone. How you go about this task depends on whether you're using an Amazon Echo or a third-party device that has Alexa built in.

Adding your Echo device

Setting up your Amazon Echo mostly means getting Alexa connected to your home's Wi-Fi network because Alexa absotively, posolutely requires access to the Internet to do its cloud-based thing. How you go about that depends on whether your Alexa device has a screen.

Adding an Echo device with a screen

If you have an Echo Show, you can use the screen to connect your device to your Wi-Fi network, which adds the device to the Alexa app automatically. Here are the steps to follow:

1. **Select your Wi-Fi network.**

   - *If you're starting your Echo for the first time,* the initial setup routine will eventually show you a list of available Wi-Fi networks. Tap the network you want to use.

   - *If your Echo is already started,* either say, "Alexa, go to Settings," or swipe down from the top of the screen and then tap Settings. On the Settings screen, tap Network and then tap the network you want to use.

   The Echo prompts you to enter the Wi-Fi network password.

2. **Use the on-screen keyboard to enter the network password, and then tap Done.**

Below the password, note that the Save Password to Amazon check box is selected. If you plan on setting up other Echo devices in the future, leave this check box selected to store your network password on Amazon's servers, which lets you skip entering the password for the other Echo devices you set up. You can always delete your saved passwords from Amazon, as I describe in Chapter 15.

3. **Tap the Connect button.**

The Echo connects to your Wi-Fi network, and then prompts you to sign in to your Amazon account.

4. **Sign in to your account, as I describe earlier.**

Your Echo Show is registered to your Amazon account.

5. **Tap Continue and run through the rest of the setup chores.**

The chores you see depend on the device but include naming the device, confirming your location (for weather and traffic), and possibly installing the latest device software. When all that's done, the Echo appears in the Alexa app. To check, choose Devices ➪ Echo & Alexa.

Adding an Echo device without a screen

If your Echo doesn't have a screen, you can use the Alexa app to add the device directly. Follow these steps to add your Echo device to the Alexa app:

1. **In the Alexa app, choose Devices ➪ Add(+) ➪ Add Device.**

The Setup screen appears and you're prompted to choose the type of device you're setting up.

2. **Tap Amazon Echo.**

The Setup screen displays a list of Echo device types.

3. **Tap the type of Echo you want to configure.**

The Alexa app asks if your device is in setup mode.

4. **If your device isn't in setup mode, press and hold down the Echo device's Action button (the button with the dot; see Chapter 3) for 5 seconds.**

TIP

How do you know if your Echo device is in setup mode? The device's light ring (see Chapter 3) should be showing a spinning, orange arc and the device itself will say "Hello. Your device is ready for setup."

5. **With your device in setup mode, tap Yes.**

The Select Your Amazon Echo screen appears, as shown in Figure 2-4. The screen shows an Available Devices heading, but really what you're seeing is a list of temporary Wi-Fi networks, one for each Echo device currently in setup mode. Each network will have a name along the lines of Echo Dot-123 or Echo-ABC.

FIGURE 2-4:
The Select Your
Amazon Echo
screen shows
your available
Alexa devices.

6. **Tap your device in the Available Devices list.**

The app displays a list of available Wi-Fi networks.

If you've set up an Alexa device previously, you may see your network listed in the Previously Saved to Amazon section.

7. **Tap your Wi-Fi network.**

The app prompts you to enter the network password.

If you've set up an Amazon device on the same network previously, and you elected to save your Wi-Fi password with Amazon, the Alexa app connects your new device to your network right away. Thanks, Amazon! You can now merrily skip to Step 9.

8. **Type the password.**

Below the password, note that the Save Password to Amazon check box is selected. If you plan on setting up other Echo devices in the future, leave this check box selected to store your network password on Amazon's servers, so you can skip entering the password for the other Echo devices you set up. You can always delete your saved passwords from Amazon, as I describe in Chapter 15.

9. **Tap Connect.**

The app connects your Alexa device to your Wi-Fi network.

10. **Tap Continue.**

The app takes you through a few more setup chores, depending on the device.

11. **Follow the on-screen instruction to finish setting up your device.**

When all that's done, the Echo appears in the Alexa app. To check, choose Devices ⇨ Echo & Alexa.

Adding a third-party Alexa device

If you have a third-party device that has Alexa built-in — such as a Sonos One speaker or an ecobee smart thermostat or switch — follow these steps to add that device to the Alexa app:

1. **Install and turn on the device.**

How you do this depends on the device, so see the little booklet that came with the device for details.

2. **Install the manufacturer's app on your mobile device.**

3. **Use the manufacturer's app to set up the device.**

For a Wi-Fi device, the setup process will get the device connected to your Wi-Fi network. For a Bluetooth device, you'll get instructions for putting the device into pairing mode (which I discuss in the upcoming "Connecting speakers or headphones via Bluetooth" section). You'll likely also have to create an account with the manufacturer, name the device, and give the manufacturer's app various permissions.

4. **When the setup procedure asks for permission to link the device to your Amazon account (see Figure 2-5), go ahead and allow it.**

The device's built-in Alexa won't work unless the device is linked to your Amazon account. Make sure you use the same Amazon account as the one you're using with the Alexa app.

5. **In the Alexa app, choose Devices ⇨ Add (+) ⇨ Add Device, tap the device category, tap the device brand (or tap Other if you don't see the brand), and then tap Discover Devices.**

TIP

Alternatively, if you have another Alexa device set up, you can say "Alexa, discover devices."

Alexa looks around for your device and then lets you know when it has been discovered.

FIGURE 2-5:
When setting up
your third-party
Alexa device, you
need to link the
device to your
Amazon account.

Add Amazon Alexa

Amazon Alexa is a free voice service that
lets you use voice commands to play
music, get the news and weather, and
more.

Add to Sonos

Personalizing Your Alexa Device

With your Alexa device up and running and connected to Wi-Fi, let the fun begin!
Feel free to skip ahead to Chapter 3 to learn all the Alexa basics you need. However, to discover a few useful tasks for personalizing Alexa and your Alexa devices,
continue reading this section.

Creating voice profiles

If you're the only person who uses Alexa, you don't have to worry about whether
Alexa recognizes you. However, if multiple people in your home share a single
Alexa device, you'll want Alexa to differentiate between each voice for shopping,
voice calling, text messaging, and so on. To get Alexa to recognize different voices,
you need to set up a *voice profile* for each person.

Fortunately, setting up a voice profile isn't complex. Begin by muting any other
Alexa devices within range. (On an Echo, for example, tap the Mute button, which
is the button with a microphone icon that has a line through it; see Chapter 3.)
Make sure there are no loud ambient noises, get within a few feet of the device,
and then say, "Alexa, learn my voice." Alexa greets you, asks your name, and then
asks you to repeat a bunch of phrases such as "Alexa, what are your deals?" and
"Amazon, search for holiday music."

Now that your voice is recognized, other people get do the same. Here are the steps each person must follow.

1. **Install the Alexa app on your mobile device and then launch the Alexa app.**

   If you prefer to perform these steps on a mobile device that already has Alexa installed (because, say, you don't have a device of your own), use that device's Alexa app to sign out (by selecting More ⇨ Settings ⇨ Sign Out) and then continue with these steps.

2. **Sign in using the main Amazon account that your family uses for shared Alexa devices.**

   The Setup screen appears.

3. **Tap I'm Someone Else.**

   The Enter Your Name screen appears.

4. **Enter your first and last name and then tap Continue.**

5. **Go through the setup screens until you see the Teach Alexa Your Voice screen.**

6. **Tap Continue and run through the steps to create a voice profile.**

Giving the device a fun name

When Alexa sets up a new device, it gives the device a name such as Paul's Echo or Madge's Echo Dot. Boring! I suppose it doesn't matter all that much what name a device uses, but if you, like me, find these default names too dull to live with, replace them! Here are the steps:

1. **In the Alexa app, choose Devices ⇨ Echo & Alexa.**

   The Echo & Alexa screen appears and displays a list of available Alexa devices. Look at all those boring names!

2. **Tap the device you want to rename.**

   The Device Settings screen appears, which includes the device name at the top.

3. **Below the device name, tap Edit Name.**

   The Alexa app opens the name for editing and displays the on-screen keyboard.

4. **Delete the old name and type the name you prefer.**

5. **Tap Done (Android) or Return (iOS or iPadOS).**

   The Alexa app updates the device with the new name.

Setting the device location

The physical address of your Alexa device is important because Alexa uses that location to provide you with weather, traffic, and a few other goodies specific to your locale. The Alexa app gets your address from your Amazon account or the location services on your mobile device or both, but you can specify a different address by following these steps:

1. **In the Alexa app, choose Devices ⇨ Echo & Alexa.**

The Echo & Alexa screen shows up with a list of available Alexa devices.

2. **Tap the device you want to work with.**

The Device Settings screen appears.

3. **Tap Device Location.**

The Alexa app displays the Location screen, which shows the current country and address for the device.

4. **To specify a different country, tap Change to the right of the current country value, and then select the country from the list provided.**

5. **For the other location fields, tap inside each field to open it for editing, and then use the on-screen keyboard to modify the value.**

6. **When you're done, tap Save.**

The Alexa app updates the device with the new location.

Setting the device time zone

Alexa is happy (if that's the right word) to tell you the time when asked or to set a reminder at a specified time, but that time is meaningful only if Alexa has the correct time zone for you. Here's how to change the time zone:

1. **In the Alexa app, choose Devices ⇨ Echo & Alexa.**

The Echo & Alexa screen appears and displays a list of available Alexa devices.

2. **Tap the device you want to work with.**

The Device Settings screen appears.

3. **Tap Time Zone.**

The Alexa app displays the Time Zone screen.

4. **Specify your region by tapping Change, selecting the region from the list provided, and then tapping Done.**

The Alexa app displays a list of time zones for the region you chose.

5. **Tap the time zone you want to use.**

The Alexa app reminds you to check any existing alarms to make sure they're still set to the correct time after this time zone change.

6. **Tap OK.**

The Alexa app updates the device with the new time zone.

Changing the device language

By default, your Alexa device uses the same language as you use with your associated Amazon account. If you prefer to use a different language (for example, you want Alexa to speak with a English accent), follow these steps to change the device language:

1. **In the Alexa app, choose Devices ⇨ Echo & Alexa.**

The Alexa Devices screen appears and displays a list of available Alexa devices.

2. **Tap the device you want to work with.**

The Device Settings screen appears.

3. **Tap Language.**

The Alexa app displays the Language screen, which displays a list of available languages.

The Language setting appears for only Amazon Echo devices, so you won't see it if you're configuring a third-party Alexa device.

4. **Select the language you want to use.**

The Alexa app warns you that changing the language setting might prevent Alexa from understanding what you say.

5. **Tap OK.**

The Alexa app updates the device with the new language.

Changing the measurement units

When you ask Alexa for the current weather or the forecast, the temperature unit Alexa uses depends on your Amazon account settings. If that unit is wrong — for example, Alexa is using Fahrenheit when you're clearly a Celsius person — you can easily change that. The same goes for distance units, for which you can choose either miles or kilometers, depending on where you fall in the scale of metric-friendliness.

Follow these steps to update your Alexa device with the units you prefer:

1. **In the Alexa app, choose Devices ⇨ Echo & Alexa.**

   The Echo & Alexa screen appears and displays a list of available Alexa devices.

2. **Tap the device you want to work with.**

   The Device Settings screen appears.

3. **Tap Measurement Units.**

   The Alexa app displays the Measurement Units screen.

4. **Under Temperature, select either Fahrenheit or Celsius.**

5. **Under Distance, select either Miles or Kilometers.**

   The Alexa app updates the device with the new measurement units.

Changing the background images on an Echo Show

If you have an Echo Show, you can specify a collection of images that appear as the device's Home screen background. Here's how:

1. **In the Alexa app, choose Devices ⇨ Echo & Alexa.**

   The Echo & Alexa screen appears and displays a list of available Alexa devices.

2. **Tap the device you want to work with.**

   The Device Settings screen appears.

3. **Tap Home Screen Background.**

   The Alexa app displays the Home Background screen, which enables you to manually select the photos you want to display or choose a photo collection.

4. **Select the photos or collection you want to use.**

   The Alexa app updates the device to display the new background images.

Connecting External Speakers to Your Alexa Device

As I show in Chapter 4, using a voice command to order up a song, music genre, or playlist is one of the pleasures of using Alexa at home. However, depending on your Alexa device, that pleasure may be diminished if the music you ask for is played through a speaker that offers only so-so sound quality. If your Alexa device doesn't deliver decent sound, or if you just want the best audio experience (within your budget, anyway), you'll want to connect external speakers to your Alexa device.

Similarly, if you'll be using your Alexa device to listen to anything — such as music, podcasts, or an audiobook — for an extended time and you don't want to disturb the people around you, you'll want to connect earbuds or headphones to your Alexa device.

Fortunately, most Alexa devices — and all Amazon Echo devices — can work with external speakers or headphones, via a direct cable connection or a wireless Bluetooth connection.

Connecting speakers or headphones using a cable

Most Alexa-enabled devices come with an Audio Output port (see Figure 2-6) that accepts a standard 3.5mm audio cable jack. On some older devices (such as early generations of the Echo and the Echo Plus), that port is hidden by a plastic cover (labeled, perplexingly, *AUX OUT*), which you need to move aside to access the port.

If you're connecting external speakers, be sure to position them at least 3 feet from the Alexa device (so the speakers don't interfere with Alexa's signals), and then plug the speakers into a power outlet and turn them on. Connect the speakers' or headphones' audio cable to the Alexa device's Audio Output port, and you're good to go.

Connecting speakers or headphones via Bluetooth

Alexa is on speaking terms (no pun intended) with Bluetooth, a wireless technology that enables you to make wireless connections to other Bluetooth-friendly devices. In particular, Alexa can connect to Bluetooth speakers or headphones, so you can either upgrade or hide the sound output of your Alexa device.

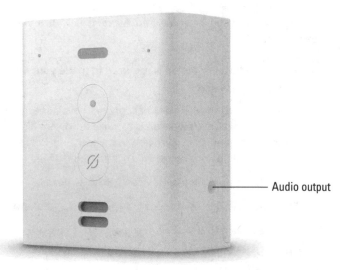

Audio output

FIGURE 2-6:
The Audio Output
port on an
Echo Flex.

Photograph courtesy of Amazon

In theory, connecting Bluetooth devices should be criminally easy: You bring them within 33 feet of each other (the maximum Bluetooth range), and they connect without further ado. In practice, there's usually a bit of further ado (and sometimes plenty of it), in one or both of the following forms:

>> **Make the devices discoverable:** Unlike Wi-Fi devices, which broadcast their signals constantly, most Bluetooth devices broadcast their availability — that is, they make themselves *discoverable* — only when you tell them to. This makes sense in many cases because you usually want to connect a Bluetooth component, such as a headset, with only a single device. By controlling when the device is discoverable, you ensure that it works only with the device you want it to.

>> **Pair Alexa and the device:** As a security precaution, many Bluetooth devices need to be *paired* with another device before the connection is established. This is why making a device discoverable is also known as putting the device into *pairing mode.*

To make things even less convenient, most Alexa devices (including all Echo devices) can pair with only one Bluetooth device at a time, so if you've previously paired your Alexa device with something else, you have to unpair that other device before continuing.

If you're connecting external Bluetooth speakers, place them at least 3 feet from the Alexa device, and then plug the speakers into a power outlet and turn them on.

Now follow these steps to pair your Alexa device with your Bluetooth speakers or headphones:

1. **Perform whatever action is required to make your Bluetooth speakers or headphones discoverable.**

   Most Bluetooth speakers and headphones have a special button that you press and hold for a few seconds to make the device discoverable. See your device's user manual to learn how to make the device discoverable (or to put the device into pairing mode, which is the same thing).

2. **In the Alexa app, choose Devices ⇨ Echo & Alexa.**

   The Echo & Alexa screen appears and displays a list of available Alexa devices.

3. **Tap the device that you want to pair with the speakers or headphones.**

   The Device Settings screen appears.

4. **Tap Bluetooth Devices.**

   The Alexa app displays a list of devices you've paired previously. This list will be empty when you're just starting out, but after you've paired a device, you can pair it again later by selecting it from this list.

TIP

   If the Bluetooth device that you want to pair with again was the most recently paired device, you can say "Alexa, connect" and Alexa will automatically pair with that device (assuming the device is on and within range).

5. **Tap Pair a New Device.**

   The Alexa app searches for nearby devices that are discoverable and then displays in the Available Devices list the name of each device it finds. Note that it may take anywhere from a few seconds to a minute or two for your device name to appear.

6. **Tap the device you want to pair.**

   The Alexa app pairs the speakers or headphones and you're Alexa devices says "Connected."

Disconnecting paired Bluetooth speakers or headphones

If you're using a paired set of Bluetooth speakers or headphones with your Alexa device, you may decide later that you prefer to listen directly from the Alexa device's internal speaker or via external speakers connected with a cable. Or you may want to pair your Alexa device with some other Bluetooth device. In any of these cases, you need to disconnect your Bluetooth speakers or headphones by following these steps:

1. **In the Alexa app, choose Devices ⇨ Echo & Alexa.**

   The Echo & Alexa screen appears and displays a list of available Alexa devices.

2. **Tap the device that has the paired speakers or headphones.**

   The device's Settings screen appears.

3. **Tap Bluetooth Devices.**

   The Alexa app displays a list of devices that you've paired previously.

4. **Tap the device that you want to disconnect, and then tap Disconnect Device.**

   The Alexa app disconnects the speakers or headphones. Alexa announces the disconnection over your Alexa device's internal speaker.

REMEMBER

If the Bluetooth Devices list contains devices that you no longer have or use, remove those devices to reduce clutter and make the list easier to navigate. Tap a device you want to get rid of, and then tap Forget Device.

Chapter **3**

Learning Alexa Basics

At this point in your relationship with Alexa, you should have the Alexa app installed on your smartphone or tablet, and the app should be connected to your Amazon account. If you're using an Alexa device, such as one of Amazon's Echo gadgets, that device should be powered up and connected to your Wi-Fi network. If you've checked off all those items in your to-do list, congratulations are in order: You're all set to start using Alexa!

In this chapter, you take the next step by learning some crucial basics not only for Alexa but also for Amazon's Echo smart speakers. From there, I take you on a test drive of Alexa's most useful and most common voice commands. Clear your throat and get ready to use Alexa!

Getting to Know Your Echo

You can have a great Alexa experience using just the Alexa app, but most folks opt for an Alexa device, because that hands-free convenience is hard to resist. And when it comes to Alexa devices, although lots of them are available, Amazon's various Echo smart speakers (which I describe briefly in Chapter 1) are by far the most popular way to invite Alexa into your home. So, before learning the basics of Alexa, it makes sense to first learn the basics of your Echo device.

Taking a closeup look at the far-field microphone

PCs have had either external or built-in microphones for a few decades, and of course smartphones and tablets have had internal microphones from day one. But the characteristic that all these microphones have in common is that they assume that the speaker is relatively close — within a few inches or, at most, a foot or two. Move much farther away and those microphones get notoriously unreliable because they have trouble distinguishing your voice from the background noise in your environment.

That sort of second-rate microphone performance just won't do in devices that not only rely on voice commands to get things done but also assume that those commands could be coming from 10, 15, or even 20 feet away — devices such as the Echo and its siblings. To get accurate and clear voice recordings, your Echo relies on a technology called the *far-field microphone,* which is optimized to distinguish a voice from the ambient room noise even when that voice is far away. Your Echo's far-field microphone uses some fancy-schmancy technology to accomplish this difficult task:

>> **Microphone array:** The Echo's microphone is actually an array of seven individual microphones, with six microphones arranged around the perimeter and the seventh microphone in the center, as shown by the circles in Figure 3-1. (Note that Figure 3-1 shows an older generation of Echo, but the microphone principle remains the same in the newest Echo speakers. Also note that in devices such as the Echo Dot, the array uses a smaller number of microphones.) Using an array of microphones in this way means you can talk to your Echo from any direction.

>> **Noise reduction:** Detects unwanted audio signals — known in the audio trade as *noise* — and reduces or eliminates them.

>> **Acoustic echo cancellation:** Detects sounds coming from a nearby loud-speaker (such sounds are known as *acoustic echo*) — even if that loudspeaker is the Echo device itself — and reduces or cancels them to ensure accurate voice recordings.

>> **Beamforming:** Uses the microphone array to determine the direction your voice is coming from and then uses that directional information to home in on your voice.

>> **Barge-in:** The microphone ignores whatever media the Alexa device is currently playing — such as a song or podcast — so that the microphone can more easily detect and recognize a simultaneous voice command (thus enabling that command to "barge in" on the playing media).

>> **Speech recognition:** Detects the audio patterns associated with speech and focuses on those patterns instead of any surrounding noises.

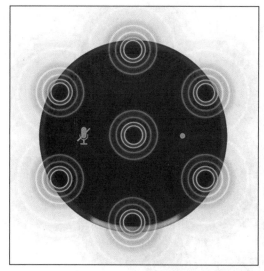

FIGURE 3-1:
Your Echo uses an array of seven microphones to hear what you're saying.

Photograph courtesy of Amazon

Pushing the Echo's buttons

The Echo is built to be a hands-free device, which is a welcome design choice when you have your hands covered in dough and want to change the music or know the time. However, *hands-free* doesn't mean *hands-off* because your Echo's outer shell is festooned with anywhere from two to four buttons that you can use to control certain aspects of the device.

Figure 3-2 shows the top of an Echo and points out the four buttons.

Here's a summary of the available buttons:

>> **Action:** Activates Alexa. That is, pressing the Action button is the same as saying Alexa's wake word.

>> **Microphone Off:** Turns off the Echo microphone. When the microphone is off, this button glows red, as does the Echo light ring. Press this button again to turn the microphone back on. On the Echo Spot and Echo Show, this button also toggles the front camera off and on.

>> **Volume Up:** Increases the volume (no surprise, there).

>> **Volume Down:** You guessed it, decreases the volume.

REMEMBER

Not every Echo device comes with all four buttons. For example, the Echo Flex doesn't have volume buttons, and the Echo Show doesn't have the Action button. Also, some older versions of the Echo and Echo Plus don't have volume buttons, but you can turn the device's light ring to adjust the volume up (clockwise) or down (counterclockwise).

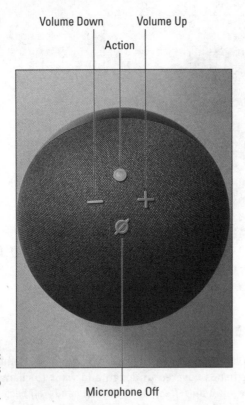

Volume Down   Volume Up

Action

Microphone Off

FIGURE 3-2:
The buttons
that dot the top
of the Echo.

Getting feedback from Alexa

When you interact with Alexa, the program almost always offers some kind of response, which at least lets you know that Alexa is listening. Following are the four main response types:

>> **Voice:** You always get a voice-based answer when you ask Alexa a question, but some other Alexa interactions also elicit a voice response (such as a simple "Okay" when you ask Alexa to turn on the lights).

>> **Visual:** On devices that have a screen, you almost always see a visual response to a voice command, such as the time if you asked Alexa what time it is, or the current volume level if you asked Alexa to adjust the volume.

>> **Sound:** Some commands rate only a simple tone or similar sound as a response. On an Echo Show, for example, asking Alexa to turn the volume up or down elicits a tone response (in addition to showing the new volume level on the screen).

>> **Light ring:** The Echo and Echo Dot come with a *light ring,* a circular light around the bottom circumference of the device. This is a surprisingly complex feature that offers no less than a *dozen* different messages depending on the colors and their movements:

- *No lights:* Alexa is waiting for your next voice command. How can you tell the difference between this mode and the device being turned off? You can't!

- *All blue with a spinning cyan arc:* The Alexa device is starting.

- *All blue with a stationary cyan arc:* Alexa is processing a voice command; the cyan arc is pointing in the direction of the person doing the talking.

- *All blue alternating with all cyan:* Alexa has submitted your voice command for processing to the online Amazon Voice Service and is waiting for the response.

- *All red:* The device microphone is turned off.

- *Pulsing all yellow:* A message or a notification is waiting for you.

- *Pulsing all green:* You have an incoming call or a Drop In (that is, someone contacting you from another Alexa device on your network; see Chapter 5).

- *Spinning green:* You're on a call or Drop In.

- *Pulsing white arc:* You're adjusting the volume. The more white you see, the louder the volume.

- *Spinning orange arc:* The device is in setup mode.

- *Pulsing all purple:* There was an error trying to connect to your Wi-Fi network.

- *Flash of all purple:* Do Not Disturb mode is activated.

Checking up on Alexa

After you issue a request or a command to Alexa, the device might respond in an unexpected way or it might perform an action that has nothing to do with what you said. These sorts of errors usually mean that Alexa misinterpreted what you said, which is almost always a temporary glitch. If you're getting a consistent error, however, you can troubleshoot the problem by checking with Alexa to see what it *thinks* you said. You do this checking by issuing either of the following requests:

>> "Alexa, tell me what you heard."

>> "Alexa, why did you do that?"

Operating the Echo touchscreen

If you have an Echo Show, you get a third way (besides issuing voice commands and pressing the top buttons) to interact with the device: letting your fingers do the walking on the device touchscreen.

You'll get comfy with this screen quickly if you're already used to a smartphone or tablet touchscreen. If you're new to a touchscreen, just realize that you use a finger to make things happen on the screen. You have two main moves to learn:

>> **Tap:** Use a finger to quickly press and release the screen where desired. This gesture is what you use to initiate just about any action on the device. The tap selects commands, toggles switches on and off, and much more.

>> **Swipe:** Drag a finger across the screen. For example, drag a finger from the top of the screen down to see the main menu, from which you can return to the Home screen (by tapping the house icon) or display the device settings (by tapping the gear icon). As another example, when you have the Home screen displayed, you can swipe right or left to see other content.

Getting to Know Alexa

Alexa is designed to be as simple as possible and to have a shallow learning curve, especially at the start. That's good news if you're just beginning with Alexa because it means you have to learn only a few basics. The next few sections walk you through what you need to know.

Getting Alexa's attention

When Alexa is not in use, it may seem as though it's just sitting there listening to you talk to yourself, but that's not the case. Instead, it's more accurate to imagine Alexa spending its off time in a light, dreamless slumber where it has no idea what's happening around it. Your job is to interrupt that slumber by gently tapping Alexa to rouse it and get its attention.

There are several ways to get Alexa's attention, but here are the most common:

>> **Say Alexa's wake word.** The default wake word is *Alexa,* but you can change it (see Chapter 15).

>> **Press the Action button (refer to Figure 3-2), if your Echo device has one.**

>> **In the Alexa app, tap the Alexa button.**

Keeping Alexa's attention

Most of the time, you interact with Alexa by issuing intermittent voice commands: You ask for the time, the temperature a minute later, and whether Dustin Hoffman was in *Star Wars* a few minutes after that. However, every now and then you may want to issue a series of commands, one after the other. That's perfectly fine, but it gets a bit old having to say, "Alexa" at the start of each voice command.

Forget that. Instead, you can put Alexa into Follow-Up mode, which enables you to say the wake word once and then issue multiple commands without having to say the wake word again.

Follow-Up mode is slick, but you should be aware that Alexa won't switch to Follow-Up mode in these situations:

>> You've forced Alexa to stop responding by issuing a command to end the conversation, which I describe in the next section.

>> Your Alexa device is playing media (such as a song or an audiobook).

>> Alexa can't be certain that you're speaking to it and not to someone else nearby.

Here are the steps to trudge through to enable Follow-Up mode:

1. **In the Alexa app, choose Devices ⇨ Echo & Alexa.**

The Echo & Alexa screen appears with a list of your Alexa devices.

2. **Tap the device you want to work with.**

The Device Settings screen appears.

3. **Tap Follow-Up mode.**

4. **Tap the Follow-Up Mode switch to on, as shown in Figure 3-3.**

The Alexa app updates the device with the new setting.

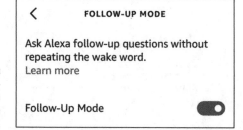

FIGURE 3-3: Tap the Follow-Up Mode switch to on to enable Follow-Up mode.

TECHNICAL STUFF

As this book went to press, Amazon was working on (but hadn't yet released) a new Alexa feature called *natural turn-taking,* which builds on follow-up mode by enabling you to converse with Alexa in a more natural way without having to repeat the wake word. Natural turn-taking might be available by the time you read this, so keep an eye out for it.

Adjusting Alexa's speaking rate

Alexa speaks at a rate that could be called conversational — not too fast, not too slow. However, you might not be happy with the speed at which Alexa talks. If you're hard of hearing, for example, Alexa might talk way too fast for you to understand; if you're always in a hurry, Alexa's speaking rate might be drive-you-crazy slow.

Fortunately, you can control Alexa's speaking rate by using the following requests:

>> "Alexa, speak faster."

>> "Alexa, speak slower."

>> "Alexa, speak at your normal rate."

TIP

You can repeat the "speak faster" or "speak slower" request to make Alexa speak even faster or slower, respectively. Eventually, Alexa reaches its limit and tells you "Sorry, I'm already at my highest (or lowest) speaking rate."

Ending the conversation

Alexa, bless its digital, cloud-based heart, can be very helpful, but sometimes it's *too* helpful. That is, you can ask Alexa a question and still be listening to the answer a minute later! If you find that Alexa is going on and on about something, you can end the monologue (Amazon calls this "ending the conversation" — ha!) by interrupting Alexa with any of the following commands:

>> "Alexa, stop."

>> "Alexa, enough."

>> "Alexa, shush."

>> "Alexa, cancel."

>> "Alexa, thank you."

>> "Alexa, sleep."

Enabling Brief mode

Alexa's infamous long-windedness has long been a complaint of users, so Amazon finally did something about it: It created a feature called Brief mode. In Brief mode, Alexa gives shorter answers than usual, and in situations where it usually gives a quick "Okay" or similarly useless response, it now just plays a sound. If that sounds like bliss to you, follow these steps to turn on Brief mode:

1. **In the Alexa app, choose More ⇨ Settings.**

2. **Tap Voice Responses.**

3. **Tap the Brief Mode switch to on, as shown in Figure 3-4.**

   The app configures Alexa to use Brief mode.

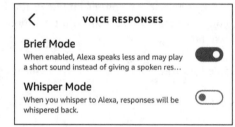

FIGURE 3-4:
Tap the Brief Mode switch to on to enable Brief mode.

Enabling Whisper mode

If you have people nearby, the last thing they probably want to hear is both your Alexa utterances and Alexa's responses at full volume. Does that mean you have to forgo Alexa until you're alone? Not necessarily. You can enable Alexa's Whisper mode, which means that when you whisper a request to Alexa, it whispers its response back. (Yes, this is hilariously weird at first, even a bit creepy. Perhaps you'll get used to it!)

The easiest way to enable Whisper mode is with a voice request:

"Alexa, turn on Whisper mode."

Inanely, Alexa responds at full volume to tell you that Whisper mode has been activated! Alternatively, you can enable Whisper mode more quietly by using the Alexa app:

1. **In the Alexa app, choose More ⇨ Settings.**

2. **Tap Voice Responses.**

3. **Tap the Whisper Mode switch (see Figure 3-4) to on.**

The app configures Alexa to use Whisper mode.

Enabling Multilingual mode

By default, Alexa speaks whatever language you've set up in your Amazon account. To check this, sign in to Amazon, select Account & Lists, and then select Language Settings. However, you're free to use a different language than the one in your account. In particular, you can enable Alexa's Multilingual mode, which lets you converse with Alexa in two languages.

The quickest way to enable Multilingual mode is to use a voice request:

"Alexa, speak *language.*"

Replace *language* with the language that you want Alexa to speak (such as Spanish or French) in addition to your default language.

Alternatively, you can enable Multilingual mode using the Alexa app:

1. **In the Alexa app, choose Devices ⇨ Echo & Alexa.**

The Echo & Alexa screen appears with a list of your Alexa devices.

2. **Tap the device you want to work with.**

The Device Settings screen appears.

3. **Tap Language.**

4. **Tap the language combo you want Alexa to use, such as English/Español or English/Français.**

   The Alexa app asks you to confirm the language change.

5. **Tap OK.**

   The Alexa app updates the device with the new language setting. (The process can take a minute or two.)

Some Useful Alexa Commands

The rest of the book is all about configuring and using Alexa for various situations, including playing music, sending messages, helping out at home, and controlling home-automation devices. I dive deep into each topic, but if you feel like wading in the shallow end for a bit longer, the rest of this chapter is just for you. Here, I take you through a few dozen of Alexa's most common and most useful voice commands. Think of what follows as Alexa 101.

Everyday info commands

Alexa excels at everyday requests for information related to the weather, news, and traffic. Here are the basic commands (see Chapters 6, 7, and 8 for many more):

>> "Alexa, what time is it?"

>> "Alexa, what time is it in *city*?"

>> "Alexa, what is today's date?"

>> "Alexa, when is *holiday* this year?"

>> "Alexa, what's the weather like?"

>> "Alexa, what's the temperature?"

>> "Alexa, will it rain today?"

>> "Alexa, what will the weather be like *day*?" (For example, "Alexa, what will the weather be like tomorrow?")

>> "Alexa, play my Flash Briefing."

>> "Alexa, what's in the news?"

- >> "Alexa, how's the traffic?"
- >> "Alexa, what was the score of yesterday's *team* game?"
- >> "Alexa, when is the next *team* game?"
- >> "Alexa, what movies are playing?"
- >> "Alexa, find me a nearby *cuisine* restaurant." (For example, "Alexa, find me a nearby Mexican restaurant.")

Information requests

Give the following commands a whirl to get general information from Alexa (see Chapter 6 to learn quite a few more):

- >> "Alexa, what's the definition of *word*?"
- >> "Alexa, how do you spell *word*?"
- >> "Alexa, what are some synonyms for *word*?"
- >> "Alexa, convert *number units1* to *units2*." (For example, "Alexa, convert 100 miles to kilometers.")
- >> "Alexa, what is *number operator number*?" (For example, "Alexa, what is 123 times 456?")
- >> Alexa, what is *number1* percent of *number2*?" (For example, "Alexa, what is 15 percent of 78.53?")
- >> "Alexa, what is the population of *city or region*?"

Audio commands

Here are the basic voice commands to try out for controlling music, podcasts, and other audio (see Chapter 4 for many more):

- >> "Alexa, play *song or album*."
- >> "Alexa, play *music genre*."
- >> "Alexa, play *playlist*."
- >> "Alexa, play the latest *artist*."
- >> "Alexa, play Sunday morning music."
- >> "Alexa, who sings this song?"

- » "Alexa, shuffle mode *on or off*."

- » "Alexa, pause."

- » "Alexa, next."

- » "Alexa, previous."

- » "Alexa, louder."

- » "Alexa, softer."

- » "Alexa, set volume to *number from 1 through 10*."

- » "Alexa, turn this off in *number* minutes."

- » "Alexa, play *audiobook title*."

- » "Alexa, read *Kindle book title*."

Video commands

Here are a few useful commands related to watching TV shows and movies (see Chapter 4 for more):

- » "Alexa, what is the IMDb rating for *TV show or movie title*?"

- » "Alexa, tell me about the movie *movie title*."

- » "Alexa, who stars in *TV show or movie title*?"

- » "Alexa, what is *actor*'s latest movie?"

- » "Alexa, how many Oscars has *actor* won?"

Alarm and timer commands

Here are the basic commands to use to set and work with alarms and timers (as I describe in Chapter 7):

- » "Alexa, set alarm for *time*."

- » "Alexa, wake me up every day at *time*."

- » "Alexa, wake me up at *time* to *music or radio station*."

- » "Alexa, snooze for *number* minutes."

- » "Alexa, set a timer for *number* minutes."

» "Alexa, set a *name* timer for *number* minutes." (For example, "Alexa, set a bread timer for 40 minutes.")

» "Alexa, how long is left on the timer?"

» "Alexa, how long is left on the *name* timer?"

» "Alexa, stop the timer."

Calendar, reminder, and list commands

Following are a few basic Alexa commands for controlling your calendar, creating reminders, and managing lists (see Chapter 7 for tons more):

» "Alexa, what's on my calendar?"

» "Alexa, what's on my calendar on *date*?"

» "Alexa, add an event to my calendar."

» "Alexa, create a new appointment."

» "Alexa, add *event* to my calendar on *day* at *time*."

» "Alexa, what am I doing tomorrow?"

» "Alexa, how many days until *date*?"

» "Alexa, how many days until *holiday*?"

» "Alexa, what time does the sun rise on *date*?"

» "Alexa, what time does the sun set on *date*?"

» "Alexa, remind me to *task* in *number* minutes."

» "Alexa, add *task* to my to-do list."

» "Alexa, what's on my to-do list?"

Communication commands

Here are the basic commands for placing calls and sending text messages (I supply lots more in Chapter 5):

» "Alexa, call *name*."

» "Alexa, call *phone number*."

» "Alexa, answer the call."

>> "Alexa, hang up."

>> "Alexa, drop in on the *room or location*."

>> "Alexa, play messages."

>> "Alexa, message *name*."

>> "Alexa, announce that *message*."

Shopping commands

Here are a few commands to place and track Amazon orders (see Chapter 6):

>> "Alexa, add *item* to my shopping list."

>> "Alexa, buy more *item*."

>> "Alexa, order *item*."

>> "Alexa, what's on my shopping list?"

>> "Alexa, where's my stuff?"

>> "Alexa, what are your deals?"

Smart-home commands

Here are some commands you can use to control various home-automation gadgets (see Chapters 11 and 12 for many more):

>> "Alexa, *device name* on."

>> "Alexa, *device name* off."

>> "Alexa, dim *device name* to *number* percent."

>> "Alexa, set *device name* to blue."

>> "Alexa, show *camera name*."

>> "Alexa, set *device name* temperature to *number* degrees."

>> "Alexa, lock *device name*."

2

Having Fun with Alexa

IN THIS PART . . .

Make the most of Alexa's media skills by finding out how to enjoy music and podcasts, listen to audiobooks, play radio stations, and watch videos.

Turn Alexa into a hands-free communications system by learning how to exchange texts, place voice and video calls, and talk directly with other Alexa users in your home.

Enjoy your home life more by using Alexa for shopping, creating lists, and getting the latest news and weather.

Make your life more efficient by using Alexa to manage your calendar, create reminders, and set alarms.

Chapter **4**

Playing Media

M ost Alexa-enabled devices are smart speakers. The *speaker* part of that moniker means the main job of those devices is to output sound. Sure, much of the time that sound is Alexa answering a question (as I talk about in Chapter 8), but more often than not, what Alexa users really want coming out of their speakers is media. What do I mean by *media*? Mostly music, but also podcasts, audiobooks, and even Kindle books, if you're into that. Where does the *smart* part of the smart speaker come into play (so to speak)? It means you can request a song, an artist, an album, a playlist, or a genre with just a voice command. And that's only the music side of things. Your favorite podcasts and other audio are also just a voice request away. And if you have an Alexa device with a screen, you can also request video content, such as TV shows and movies.

In this chapter, you explore Alexa's multifaceted multimedia prowess. You learn how to tell Alexa which music service you use, how to request and control music, how to tune into a radio station, how to order up a podcast or audiobook, how to watch videos, and much more.

Listening to Music on Alexa

If you're cooking, eating a meal with your family, or in the middle of a rousing game of Twister, and you get an urge to hear a particular song or artist, the last thing you want to do is interrupt what you're doing to wrestle with a music app or root around through your music collection to find what you want. Fortunately, with Alexa in the house, those days are long gone because now you can issue a voice command to order up whatever music you feel like hearing, all without moving an inch from where you are. Playing music is one of the most popular Alexa skills, and the next few sections tell you all you need to know.

Playing music through Alexa

The first thing you have to decide is how you want to play your music through Alexa. You have two choices:

>> **Use a music service provider.** A music service provider is a third-party service that supplies you with music, sometimes for a fee. Examples include Amazon Music, Apple Music, and Spotify. If this is the route you want to take, see "Linking to a music provider," next.

>> **Use your mobile device music.** If your music is stored on a smartphone or tablet, you can stream that music through your Alexa device by using a Bluetooth connection. To give this technique a whirl, see "Streaming Mobile Device Audio through Alexa," later in this chapter.

Linking to a music provider

Alexa doesn't have access to music on its own. Instead, you need to connect it to whatever music provider service you use. The services supported by Alexa depend on where you live, but common providers include Amazon Music, Apple Music, Spotify, and TuneIn.

REMEMBER

Some music providers require you to have a special type of account or subscription before they'll let you connect to Alexa. For example, you can connect Spotify to Alexa only if you have a Spotify Premium subscription.

Before Alexa can use a music provider, you need to give Alexa permission to access your account on that provider. This is called *linking* your account, and you need to link each music provider you want to access through Alexa. (The exception here is Amazon Music, which is automatically linked to Alexa when you log in to Amazon using the Alexa app.) Follow these steps to link a music provider to Alexa:

1. **In the Alexa app, choose More ⇨ Settings.**

   The Settings menu appears.

2. **Tap Music & Podcasts.**

   The Music & Podcasts screen appears, listing the music providers currently linked to Alexa.

3. **Tap Link New Service.**

   You see the Link Service screen, which lists the available music providers you can add.

4. **Tap the service to which you want to link.**

   The Alexa app displays the skill page for the service. Figure 4-1 shows the Spotify skill page. A *skill* is a special add-on that gives Alexa new features; see Chapter 9 for lots of skill-related info.

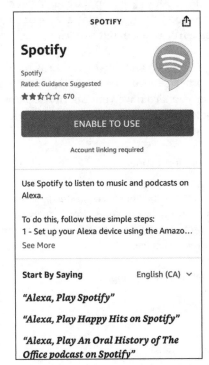

FIGURE 4-1:
Music providers (such as Spotify shown here) are linked to Alexa using the provider's skill.

5. **Tap Enable to Use and then follow the instructions that appear.**

   In most cases, you need to log in to your music service and then tap the link or button that authorizes Alexa to access your account.

6. **When the link is complete, tap Close.**

   The Alexa app displays a screen to let you know that you can use Settings to change the default music provider.

7. **If you want to change the default provider, tap Visit Music Settings and then follow the instructions in the next section; otherwise, tap Not Now.**

   The linked music provider is now ready to use with Alexa.

Setting up a default music provider

When you connect the Alexa app to your Amazon account, you get automatic access to Amazon Music, which Alexa sets up as your default music provider. Here, *default* means that when you ask Alexa to play a song or an artist, for example, Alexa retrieves that music from the default service.

Note, too, that just because you set up one music service as the default, it doesn't mean that you can't use another service. For example, if you are using Amazon Music as the default provider you also have your Spotify account linked to Alexa, you can still play something from Spotify by including the service name as part of your voice command. For example:

> "Alexa, play Camera Obscura from Spotify."

If you're cool with using Amazon Music as your default provider, there's nothing to see here, so feel free to move on to the next section. However, if you want to use a different provider as the default, you need to follow these steps to set up that provider as Alexa's default music service:

1. **In the Alexa app, choose More ➪ Settings.**

   The Settings menu appears.

2. **Tap Music & Podcasts.**

   The Music & Podcasts screen appears.

3. **Tap Default Services.**

   The Alexa app displays the Default Services screen, as shown in Figure 4-2. You can choose three defaults: one for general music requests; one for more music stations, such as those based on an artist or genre; and one for podcasts.

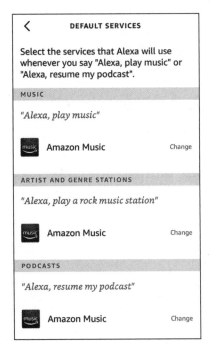

FIGURE 4-2:
Use this screen to set your music defaults.

4. **In the Music section, tap Change and then tap the music service you want to use for general music requests.**

5. **In the Artist and Genre Stations section, tap Change and then tap the music service you want to use for station requests.**

6. **In the Podcasts section, tap Change and then tap the music service you want to use for podcast requests.**

The Alexa app updates your default music settings.

Voice commands for controlling music

Okay, now it's time to get down, get funky, or get wherever you end up when you listen to your favorite music. Alexa has an extremely long list of voice commands for playing and controlling music, and the next few sections take you through them all.

Some notes to bear in mind as you work through these commands:

» **Not all these commands will work on all music services.** Almost all work as advertised on Amazon Music, but your results will vary if you try them with another provider (such as Spotify).

>> **Except where noted, all these commands work on the default music service.** To run a command on another service that you've linked to Alexa, follow the command with "on *service*," where *service* is the name of the music provider (for example, "on Spotify").

>> **You can replace the word *song* with *track* or *tune*.**

>> **You can replace the word *music* with *songs*.**

Playing music by song, album, or artist

Here are some commands to order up music by song title, album title, or artist name:

>> "Alexa, play the song *title*."

>> "Alexa, play the song *title* by *artist*."

>> "Alexa, play the album *title*."

>> "Alexa, play music by *artist*."

>> "Alexa, play popular songs by *artist*."

>> "Alexa, play the latest *artist* album."

>> "Alexa, play that song that goes *lyrics*." (For example, "Alexa, play that song that goes 'Who let the dogs out.'")

>> "Alexa, play the song I just bought."

Playing music by genre

To get Alexa to play songs from a particular music genre, use the following command:

"Alexa, play *genre* music."

For *genre*, there are dozens of possibilities, but here are the most popular:

Acoustic	Electronic	Metal
Alternative	Folk	New age
Blues	Gospel	Opera
Broadway	Hard rock	Pop
Children's	Hip hop	R&B

Christian	Holiday	Rap
Classic rock	Indie	Rock
Classical	International	Soundtracks
Country	Jazz	Vocal
Dance	Latin	World
Easy listening	Meditation	

Playing music by mood or activity or both

One of Amazon Music's most interesting features lets you request music to suit a mood (such as mellow or upbeat) or an activity (such as relaxing or running) or both. Although no definitive list exists, Amazon claims that it supports more than 500 moods and activities. Here are some commands you can use:

>> "Alexa, play *mood* music."

>> "Alexa, play *activity* music."

>> "Alexa, play music for *activity*."

>> "Alexa, play *mood activity* music."

For the mood, you can try just about any mental state, including angry, chill, energetic, feeling good, happy, joyful, laid back, mellow, relaxed, romantic, sad, or upbeat.

For the activity, try describing just about anything you're currently doing, including cleaning, cooking, eating dinner, going to sleep, meditating, partying, reading, relaxing, running, studying, waking up, working, or working out.

REMEMBER

This feature generally works only on Amazon Music. Also, not every mood, activity, or mood-and-activity combo works. Alexa tries to match your request with an existing playlist that's been created by Amazon, and if no such playlist exists — for example, if you ask for "energetic going-to-sleep music" — Alexa will tell you it can't find any music to satisfy your request.

Playing and controlling playlists

Alexa offers a few commands for creating, populating, and playing custom collections of songs, otherwise known as *playlists:*

>> "Alexa, create a playlist named *title*."

>> "Alexa, add this song to a new playlist named *title*."

>> "Alexa, add this song to the playlist named *title*."

>> "Alexa, play my playlist named *title*."

>> "Alexa, shuffle my playlist named *title*."

Getting information about music

If you want to know information about a song, album, or artist, Alexa can help. Here are a few commands to try:

>> "Alexa, who sings this song?"

>> "Alexa, who sings the song *title*?"

>> "Alexa, who is in the band *artist*?"

>> "Alexa, what year did *artist* release *song or album*?"

Discovering new music

If you're in the mood for something new, here are a few commands you can use to listen to music that's either recently released or different from what you normally play:

>> "Alexa, play new music."

>> "Alexa, play new music by *artist*."

>> "Alexa, play new *genre* music."

>> "Alexa, play the song of the day."

>> "Alexa, play *artist* station."

>> "Alexa, play some music."

>> "Alexa, play more songs like this."

>> "Alexa, play songs similar to this."

>> "Alexa, play songs similar to *title*."

>> "Alexa, play songs similar to *artist*."

>> "Alexa, play some other music I like."

>> "Alexa, play songs I haven't heard."

>> "Alexa, play *artist* songs I haven't heard."

Rating music

Alexa can make better recommendations if you tell it which songs you like and which ones you dislike. Here are some commands that'll help:

>> "Alexa, thumbs up."

>> "Alexa, I like this song."

>> "Alexa, thumbs down."

>> "Alexa, I don't like this song."

Playing popular music

If you want to hear the most popular songs, here are some commands that will get you there:

>> "Alexa, play the top songs."

>> "Alexa, play the top *genre* songs."

>> "Alexa, play the top songs from the *decade*."

>> "Alexa, play the top songs in *country*."

Controlling the volume

To get the music volume just right, here are some commands you can use:

>> "Alexa, volume up."

>> "Alexa, increase the volume."

>> "Alexa, raise the volume."

>> "Alexa, louder."

>> "Alexa, volume down."

>> "Alexa, decrease the volume."

>> "Alexa, lower the volume."

>> "Alexa, softer."

>> "Alexa, what volume level is this?"

>> "Alexa, set the volume to *number from 1 through 10*."

>> "Alexa, volume *number from 1 through 10*."

- » "Alexa, mute."
- » "Alexa, unmute."

Controlling playback

While Alexa is playing music, you can use the following commands to control the playback:

- » "Alexa, stop."
- » "Alexa, pause."
- » "Alexa, play."
- » "Alexa, resume."
- » "Alexa, next."
- » "Alexa, previous."
- » "Alexa, turn shuffle on."
- » "Alexa, turn shuffle off."
- » "Alexa, turn repeat on."
- » "Alexa, turn repeat off."
- » "Alexa, skip back *number* seconds."
- » "Alexa, skip forward *number* seconds."
- » "Alexa, restart song."
- » "Alexa, restart album."
- » "Alexa, restart playlist."

Accessing Alexa via the Amazon Music app

If you get your jams through Amazon Music, not only can you access your music via any Alexa device, but you can also access Alexa via the Amazon Music app, which is available for iOS and Android. This means you can use all the voice commands from the previous section to control the Amazon Music app. Here are the steps to follow to get started:

1. **Install and then start the Amazon Music app.**

   The app asks you to sign in to your Amazon account.

2. **Type your Amazon email address (or mobile phone number, if you have a mobile Amazon account), type your Amazon password, and then tap Sign In.**

3. **If you have two-step verification turned on, enter the code you were sent and then tap Sign In.**

4. **Tap the Alexa icon at the top of the Home screen.**

   The first time you access Alexa, the app asks your permission to use your mobile device's microphone.

5. **Tap Allow Microphone Access.**

   The Android version of the app get microphone access right away, but the iOS version requires another step.

6. **In the iOS version of the app, tap Settings and then tap the Microphone switch to on.**

   The Amazon Music app is now ready to accept voice commands. Say "Alexa" or tap the Alexa icon, and then say your command.

If you want, you can play Amazon Music content through your Alexa device. Here are the steps to follow:

1. **Tap Settings (the gear icon) in the upper-right corner of the Amazon Music app.**

2. **Tap Connect to a Device.**

   The Amazon Music app displays a list of available devices, which includes any Alexa devices on your network, as shown in Figure 4-3.

3. **Tap the Alexa device you want to use for playback.**

   Your Amazon Music content now plays through your Alexa device.

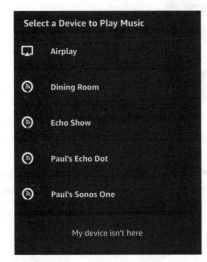

FIGURE 4-3:
Use the Select a Device to Play Music screen to play Amazon Music through an Alexa device.

The downside to playing your Amazon Music selections through an Alexa device is that you forgo the ability to use Alexa in the Amazon Music app. To get Alexa back on the job in the Amazon Music app, you must disconnect from the Alexa device. To do that, choose Settings ⇨ Connect to a Device, and then tap the Disconnect button next to your Alexa device.

Playing third-party music through Alexa

If you have a third-party music provider linked to Alexa, you normally ask Alexa to play music from that provider either by specifying the provider name in the voice command (for example, "Alexa, play the Chill playlist on Spotify") or by setting up that provider as the default, as I describe earlier (see "Setting up a default music provider").

However, a third method is available: You can use the music provider's app to select your Alexa device as the playback device. For example, in the Spotify app, if you choose Settings ⇨ Devices ⇨ Devices Menu, you see a screen offering a list of available devices that you can use for playback. This list will look similar to the one shown in Figure 4-4, which has a couple of Echo devices and a Sonos speaker. Tap the Alexa device you want to use, and any music you crank up in the app will play through your Alexa device.

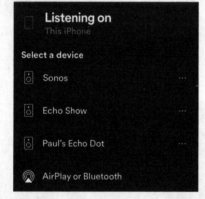

FIGURE 4-4:
The Spotify app screen, which in this example contains several devices that can be used for playback.

Adjusting music quality

Amazon Echo devices come with a feature sure to warm the cockles of audio-geeks' hearts everywhere: an equalizer. (If you're not an audio aficionado, you might not know that an *equalizer* is a device for adjusting the sound quality by controlling different sound components.) Real-world equalizers bristle with obscure settings, but Echo devices mercifully support only three:

>> **Bass:** Controls the lowest tones in the music

>> **Treble:** Controls the highest tones in the music

>> **Midrange:** Controls all tones in the music that lie between the bass and treble tones

You can use Alexa to control each of these settings as follows:

>> "Alexa, increase the bass."

>> "Alexa, increase the midrange."

>> "Alexa, increase the treble."

>> "Alexa, decrease the bass."

>> "Alexa, decrease the midrange."

>> "Alexa, decrease the treble."

>> "Alexa, set the bass to maximum."

>> "Alexa, set the midrange to maximum."

>> "Alexa, set the treble to maximum."

>> "Alexa, set the bass to minimum."

>> "Alexa, set the midrange to minimum."

>> "Alexa, set the treble to minimum."

>> "Alexa, set the bass/midrange/treble to *number from –6 through 6.*"

>> "Alexa, reset the equalizer."

TIP

If you happen to have the Alexa app open, you can also use it to make equalizer adjustments. Choose Devices ⇨ Echo & Alexa, tap the Echo device you want to adjust, and then tap Audio Settings. Use the Bass, Midrange, and Treble sliders to adjust the settings as needed.

TIP

If you have an Echo Show, you can mess up, er, sorry, I mean *mess around with* the sound quality by opening Settings and then choosing Sounds ⇨ Equalizer. Use the Bass, Mid, and Treble sliders to adjust the settings to taste.

Listening to Other Audio Sources on Alexa

One of the features that puts the *smart* in *smart speaker* is the device's capability of playing more than just music. You can also tune in to radio stations, listen to podcasts, hear audiobooks, and even have a Kindle book read to you. Now *that's* smart! The next few sections provide the details on these audio extras.

Tuning in to a radio station

Listening to individual songs, albums, artists, and playlists is great fun, but sometimes you may want someone else to do all the work. To that end, Amazon Music has a huge number of *stations* that play music from a particular genre or era. However, you may prefer to keep it old school and listen to an honest-to-goodness radio station with DJs, traffic reports, news breaks, and all the other bric-a-brac associated with live radio.

Easy money! Alexa comes with built-in support for TuneIn, which is a streaming service that offers access to live audio over the Internet. You can pay TuneIn a fee to access live sports and music events, but for our purposes you can also use TuneIn free to access hundreds of radio stations, from your local morning zoo to stations all over the world. You can get Alexa to play a radio station either by using a voice command or by using the Alexa app.

Ordering up a radio station with a voice command

To get Alexa to play a radio station hands-free, use either of the following voice commands:

>> "Alexa, play the station *frequency* on TuneIn."

>> "Alexa, play the station *call letters* on TuneIn."

>> "Alexa, play the station *name* on TuneIn."

Some notes:

>> For a local station, replace *frequency* with the station frequency, such as 96.3 or 102.1.

>> For all stations, replace *call letters* with the three- or four-letter designation that's unique to the station you want, such as WNED or CBC. If Alexa doesn't find the station, you might need to add AM or FM to the call letters.

>> For all stations, replace *name* with the station's name or nickname, such as Minnesota Public Radio or The Wolf.

>> In most cases, you probably don't need to add the "on TuneIn" part. However, if you find that Alexa is playing an Amazon Music station instead of the radio station you want, you'll need to add "on TuneIn" to your request.

Selecting a radio station with the Alexa app

Using voice commands to dial up a radio station is fast and easy *provided* you know which station you want — that is, you need to know the station's frequency, call letters, or name. If you don't have that information, or if you just want to try a radio station that specializes in a particular genre or that's based in some other part of the world, are you out of luck?

Nope, not even close. The Alexa app comes to your rescue by offering a simple interface to the TuneIn service that enables you to select the station you want. You have access to your local stations (based on your location), or you can select a station by genre (such as music or talk), location, or language.

REMEMBER

TuneIn isn't the only Alexa radio station game in town. You can add to Alexa lots of other third-party skills — both streaming services similar to TuneIn and skills created by individual radio stations — to take your radio listening up a notch or three. Turn to Chapter 9 to learn more.

Follow these steps to select a radio station using the Alexa app:

1. **In the Alexa app, choose Play.**

The Entertainment screen appears.

2. **Scroll down to the TuneIn Local Radio section and then tap that section's Browse link.**

The Alexa app displays the TuneIn browse screen, as shown in Figure 4-5.

3. **Use one of the following methods to select a station:**

- (iOS and iPadOS only) Use the Search TuneIn box to enter a search term, tap Search, and then tap the station you want to hear.

- In the Recently Played section, tap a recent station.

- Scroll down to the category you want to peruse (such as Local Radio or Top News Stations), tap the See All link or scroll the category right and left, and then tap the station you want to hear.

The Alexa app displays the Play On list (see Figure 4-6) so you can choose which Alexa device you want to use for playback.

4. **Tap the device you want to use to play the radio station.**

Alexa plays the station through the device you chose. Note that the Alexa app also displays the Now Playing strip near the bottom of the screen (refer to Figure 4-6).

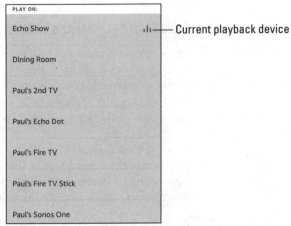

FIGURE 4-5:
The entrance to radio heaven: the TuneIn browse screen.

FIGURE 4-6:
Choose which Alexa device you want to use for the radio station playback.

Listening to a podcast

Alexa uses Amazon Music as its default source for podcasts, and while Amazon Music offers a large collection of podcasts, you might already have a collection on Apple podcasts or Spotify podcasts. If so, that's not a problem because you can link your Apple or Spotify account to Alexa (see "Linking to a music provider," earlier in this chapter) and then set that provider has your default for podcasts (see "Setting up a default music provider," earlier).

With your podcast provider linked to Alexa, you can then use the following voice commands to listen to podcasts:

>> "Alexa, play the podcast *podcast name*."

>> "Alexa, play the podcast *podcast name* on *podcast service*."

>> "Alexa, play the latest episode of the podcast *podcast name*."

>> "Alexa, play yesterday's episode of the podcast *podcast name*."

>> "Alexa, play the *date* episode of the podcast *podcast name*."

>> "Alexa, next episode."

>> "Alexa, previous episode."

- » "Alexa, pause."

- » "Alexa, resume."

- » "Alexa, fast forward *number* minutes."

- » "Alexa, rewind *number* minutes."

- » "Alexa, restart."

- » "Alexa, what podcast is this?"

TIP

If you're a podcast junkie, you'll likely be disappointed in Alexa's default podcast service. Fortunately, a fantastic Alexa skill called AnyPod enables you to subscribe to podcasts, as well as play and control any episode. Turn to Chapter 9 to learn more.

Hearing an audiobook

Music nourishes the soul, but if it's food for thought you're craving, books are what you need to consume. Most book lovers prefer the tangible quality of a print book, but in lots of situations a hardcover or paperback isn't the right choice (for example, when you're cooking a meal, exercising, or standing up and holding on for dear life on a subway or train). Also, a print book isn't shareable in the sense that two (or more) people can't read the same copy at the same time.

When you crave a book, but a print book isn't the right choice, the next best thing is an audiobook. Fortunately, Alexa's default audiobook service is Audible, a company owned and operated by Amazon. You can purchase Audible titles through Amazon, or you can sign in to Audible (www.audible.com) using your Amazon credentials and then make your purchase.

After you've purchased one or more audiobooks, they're added to your Audible library, and you can access that library via Alexa by using any of the following voice commands:

- » "Alexa, play *title* on Audible."

- » "Alexa, read *title*."

- » "Alexa, play the book *title*."

- » "Alexa, play the audiobook *title*."

If you have an Alexa device with a screen, you can use the following voice command to see a list of the books in your Audible library:

"Alexa, show me my audiobooks."

Swipe left or right to find the audiobook you want to hear; then tap the audiobook to begin the playback.

You can also use the following voice commands to control the playback:

>> "Alexa, read faster."

>> "Alexa, read slower."

>> "Alexa, pause."

>> "Alexa, resume" (if the book is now playing).

>> "Alexa, resume my book" (if some other media is now playing).

>> "Alexa, go back" (rewinds 30 seconds).

>> "Alexa, go forward" (fast-forwards 30 seconds).

>> "Alexa, restart" (restarts the current chapter).

>> "Alexa, next chapter."

>> "Alexa, previous chapter."

>> "Alexa, go to Chapter *number*."

You can also start up an audiobook using the Alexa app. Here are the steps to follow:

1. **In the Alexa app, choose Play.**

   The Entertainment screen appears. In the Audible Library section, the Alexa app displays a list of the audiobooks in your Audible library.

2. **Tap the audiobook you want to hear.**

   The Alexa app prompts you to select a playback device.

3. **Tap the Alexa device on which you want to hear the audiobook.**

   Alexa streams the audiobook through the device you chose. The Alexa app displays the audiobook in the Now Playing area (labeled in Figure 4-5).

TIP

You can use the Now Playing area to jump to another chapter. Tap the Now Playing area, tap the chapter list icon (labeled in Figure 4-7), and then tap the chapter you want to hear.

FIGURE 4-7:
After you tap
Now Playing,
you see the
audiobook
screen.

Having a Kindle book read out loud

Not only is hearing an audiobook (as I describe in the preceding section) educational or entertaining (or both), but there's also something soothing about having a good story read out loud to you. That pleasant experience is usually the case with audiobooks because they almost always hire professional voice actors as narrators. On those rare occasions when a nonprofessional (usually the author) does the reading, the results can range from not too bad to really, *really* bad.

When it comes to having Alexa read aloud the text of a Kindle book, the results fall somewhere in the middle. Instead of a professional voice actor, you get Alexa's synthesized voice reading the text. The result is listenable, although of course you don't get the changes in tone and pacing that a professional voice actor brings to the text.

You can use any of following voice commands to get Alexa to read a Kindle book out loud:

>> "Alexa, play *title* on Kindle."

>> "Alexa, read *title*."

>> "Alexa, read the book *title*."

>> "Alexa, read the Kindle book *title*."

If you have an Alexa device with a screen, you can use the following voice command to see a list of the books in your Kindle library:

"Alexa, show me my Kindle books."

Swipe left or right to find the book you want read; then tap the e-book to begin the playback.

You can use the following voice commands to control the playback:

- "Alexa, pause."
- "Alexa, resume."
- "Alexa, go back" (rewinds 30 seconds).
- "Alexa, go forward" (fast-forwards 30 seconds).
- "Alexa, next chapter."
- "Alexa, previous chapter."

You can also crank up a Kindle book reading using the Alexa app. Here are the steps to follow:

1. **In the Alexa app, choose Play.**

   The Entertainment screen appears. In the Kindle Library section, the Alexa app display a list of the books in your Kindle library.

2. **Tap the Kindle book you want to hear.**

   The Alexa app prompts you to select a playback device.

3. **Tap the Alexa device on which you want to hear the e-book.**

   Alexa streams the Kindle book through the device you chose. The Alexa app displays the Kindle book in the Now Playing area.

TIP

You can use the Now Playing area to select a different chapter. Tap the Now Playing area, tap the chapter list icon in the bottom-left corner of the screen (labeled in Figure 4-7), and then tap the chapter you want to hear.

Streaming Mobile Device Audio through Alexa

If you have music, podcasts, or other audio content on your smartphone or tablet, you may want people nearby to also hear that content or you may want to play that content through some decent speakers. You can do this in many ways, but you may be surprised to learn that one way is to stream your mobile–device audio through your Alexa device.

Yep, that *is* pretty cool and it's all done via Bluetooth, as the following steps show:

1. **In the Alexa app, choose Devices ⇨ Echo & Alexa.**

   The Echo & Alexa screen appears, displaying a list of available Alexa devices.

2. **Tap the Alexa device you want to use to stream your mobile-device audio.**

   The Device Settings screen appears.

3. **Tap Bluetooth Devices.**

4. **Tap Pair a New Device.**

   Your Alexa device is now in Pairing mode.

5. **On your mobile device, open the Bluetooth settings.**

   Choose Settings ⇨ Connections (Android only) ⇨ Bluetooth, and then make sure the Bluetooth switch is on.

   After a short wait, you should see your Alexa device in the Available Devices list (with a name similar to Echo-123), as shown in Figure 4-8.

6. **Tap the device you want to pair.**

   Your mobile device pairs with your Alexa device and says "Now connected to Bluetooth."

With the pairing complete, any audio you launch on your mobile device will play through your Alexa device. You're welcome!

TIP

After you've paired your mobile device with Alexa once, future pairings with the same device are as easy as saying, "Alexa, connect my phone."

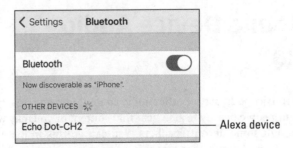

Watching Video on Alexa

If you have an Echo Show, videos are part of your media smorgasbord. Movie trailers? Check. Web videos? Check. Amazon Prime Video TV shows and movies? Check and check. The next few sections provide the details in glorious black and white.

Watching a movie trailer

If you want help deciding whether an upcoming or recently released movie is something you'd like to check out, or if you just get a kick out of previews, you can ask Alexa to play you the trailer for a movie. Use either of the following voice commands:

>> "Alexa, play the trailer for *movie title*."

>> "Alexa, show the trailer for *movie title*."

Locating and playing web videos

Alexa can search the web for videos related to a subject you specify. (Alternatively, you can leave out the subject to see a list of the currently most popular videos.) Alexa enlists the Bing search engine to scour not only YouTube but also other popular video sites such as Vimeo and Dailymotion. To run a video search, use the following voice command:

>> "Alexa, show me *subject* videos."

>> "Alexa, show me videos."

TIP

By default, Alexa searches for web videos with the Safe Search feature turned on, which means it blocks mature content. To turn off Safe Search (my, aren't *you* brave!), swipe down from the top of the Echo Show screen, choose Settings ⇨ Restrict Access ⇨ Web Video Search, and then tap the Web Video SafeSearch switch to off.

Alexa then displays a grid for the videos it found, which you can navigate by scrolling vertically. When you see a video you want to view, tap it to start the playback.

You can use the following voice commands to control the playback:

» "Alexa, pause."

» "Alexa, resume."

» "Alexa, next video."

» "Alexa, previous video."

» "Alexa, full-screen."

Watching Amazon's Prime Video

If you're an Amazon Prime member, you have access to thousands of movies and TV shows via Amazon's Prime Video service.

First, you can use any of the following voice commands to search for shows on Prime Video:

» "Alexa, show me the movie *movie title*."

» "Alexa, show me the TV show *series title*."

» "Alexa, show me new movies."

» "Alexa, show me new TV shows."

» "Alexa, show me *actor* movies."

» "Alexa, show me *genre* movies."

» "Alexa, show me *genre* TV shows."

» "Alexa, show me my video library."

» "Alexa, show me my watchlist."

Alexa displays a grid with the results from Prime Video. Scroll vertically until you see the TV show or movie you want, and then tap it to start the playback.

You can also use the following voice commands to control the playback:

» "Alexa, pause."

» "Alexa, resume."

» "Alexa, rewind" (rewinds 10 seconds).

» "Alexa, fast-forward" (fast-forwards 10 seconds).

» "Alexa, next episode."

» "Alexa, previous episode."

Watching TV with Alexa

Alexa can do an amazing range of things, but the main concern of dedicated TV watchers is using Alexa to locate, play, and control the playback of movies and TV shows on their smart TV. In the next few sections, I go through a few ways to use voice requests to interact with movies and TV shows.

Connecting your Alexa device to your TV

To set up your Alexa device to control your smart TV, you have to introduce them to each other. Here's how it's done:

1. **In the Alexa app, tap More ⇨ Settings.**

2. **Tap TV & Video.**

3. **Tap your TV type (such as Fire TV) or TV provider (such as Tubi).**

4. **Tap Link Your Alexa Device.**

   The Alexa app displays a list of your TV devices.

5. **Select the device you want to control (see Figure 4-9), and then tap Continue.**

   The Alexa app displays a list of your Alexa devices.

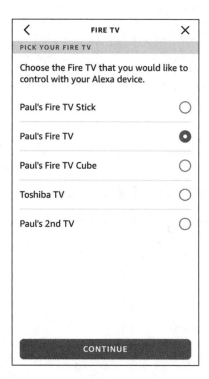

FIGURE 4-9:
Tap the device
you want to
control with
Alexa.

6. **Select the Alexa device you want to use to control the TV device you chose in Step 5.**

   If you have multiple Alexa devices, tap each device that you want to allow to control your TV device.

7. **Tap Link Devices.**

   The Alexa app connects your Alexa device and your TV device.

The way you interact with your TV depends on the device. To give you a flavor of how you can use Alexa to control your TV, the next few sections show you how to control a Fire TV device.

Navigating Fire TV tabs

To get around the Fire TV interface, you can use these voice requests:

>> "Alexa, go Home."

>> "Alexa, go to Live."

>> "Alexa, go to Your Videos" (or "My Videos" or "Videos").

>> "Alexa, go to DVR" (only if you have Fire TV Recast installed).

>> "Alexa, go to Movies."

>> "Alexa, go to TV Shows."

>> "Alexa, go to Apps."

>> "Alexa, go to Settings."

Locating movies and TV shows

You can use any of the following voice requests to search for shows on Fire TV:

>> "Alexa, show me the movie *movie title*."

>> "Alexa, show me the TV show *series title*."

>> "Alexa, show me *actor* movies."

>> "Alexa, show me *genre* movies."

>> "Alexa, show me *genre* TV shows."

In each case, you can replace "Show me" with either "Search for" or "Find."

Playing movies and TV shows

After you have a movie or TV show playing, you can use the following voice requests to control the playback:

>> "Alexa, pause."

>> "Alexa, resume" (or "Play").

>> "Alexa, rewind" (rewinds 10 seconds).

>> "Alexa, rewind *X* seconds."

>> "Alexa, fast-forward" (fast-forwards 10 seconds).

>> "Alexa, fast-forward *X* seconds."

>> "Alexa, next episode."

>> "Alexa, previous episode."

Watching live TV

If your Fire TV includes live TV channels via an HDTV antenna connection or other source, you can use the following voice requests to display the channel guide:

>> "Alexa, channel guide."

>> "Alexa, open channel guide."

>> "Alexa, show me the channel guide."

You can also use the following voice requests to tune to a station from anywhere in the Fire TV interface:

>> "Alexa, change to *channel* or *network*."

>> "Alexa, go to *channel* or *network*."

>> "Alexa, tune to *channel* or *network*."

>> "Alexa, watch *channel* or *network*."

In these requests, replace *channel* with the station's channel number (for example, "Alexa, tune to 5.1") or replace *network* with the station's network name (for example, "Alexa, watch NBC").

How do you know the station's channel number? You have three ways to find it:

>> Choose Settings ⇨ Live TV ⇨ Channel Management, and then choose your live TV source (such as Antenna Channels). In the list of channels that appears, each station displays its channel number.

>> On your Fire TV Edition version of the Alexa Voice Remote, press the Guide button to display the channel guide. As you scroll vertically through the guide, you see the channel number for each station below the station logo.

>> Use the TV Fool website (www.tvfool.com) to search for broadcast stations in your area, as I describe in Chapter 10. In the TV Fool results, the channel numbers you want are listed in the Channel section's (Virt) column.

Controlling the volume

To get the playback volume just right, here are some requests you can use:

>> "Alexa, volume up."

>> "Alexa, increase the volume."

- » "Alexa, raise the volume."

- » "Alexa, louder."

- » "Alexa, volume down."

- » "Alexa, decrease the volume."

- » "Alexa, lower the volume."

- » "Alexa, softer."

- » "Alexa, set volume to *X*" (where *X* is the volume setting you want).

- » "Alexa, volume *X*" (where *X* is the volume setting you want).

- » "Alexa, mute."

- » "Alexa, unmute."

Taking Photos with Alexa

Your Echo Show comes with a front-mounted camera. You normally use the camera for video calls, but you can also take a photo that gets stored online in Photo Booth, which is part of Amazon Photos. All Amazon accounts get 5GB of photo storage, but Amazon Prime members get unlimited photo storage, so you can take as many silly photos as you want.

To take a photo, use either of the following voice commands:

- » "Alexa, take a photo."
- » "Alexa, take a picture."

Alexa gives you a 3-second countdown (say, "Cheese!"). During that countdown, you can swipe left to add a *sticker* to the photo, which is a fun overlay such as a party hat or moustache. When the countdown ends, Alexa snaps the photo and then saves it to Photo Booth.

To view your photos, use the following voice command:

"Alexa, show my photos."

Alexa displays a slideshow of your photos. To pause the show, say, "Alexa, pause." You can then control the photos by saying "Alexa, next," "Alexa, previous," or "Alexa, resume."

Chapter **5**

Communicating with Alexa

When you think about Alexa (you *do* think about Alexa, don't you?), you probably think about standard Alexa tasks such as getting answers to your questions, summoning some mood music, and turning lights on and off. What most folks do *not* think about when Alexa comes to mind is communicating. Yep, sure, you "communicate" with Alexa in a way, but I'm talking here about communicating with other sentient beings: your family, friends, perhaps even a colleague or two.

If you have an eyebrow raised in surprise, I can't say that I blame you. Alexa's communication features are unexpected, but they're real. For example, you can send and receive text messages between any two Alexa devices or between any two mobile devices that have the Alexa app. You can also make and receive phone calls, make and receive video calls, and even use your Alexa device as an intercom.

It's all quite fun and, best of all, free (although, as usual, some restrictions apply). This chapter takes you through everything you need to know to turn your Alexa device or app into a communications system.

Managing Your Alexa Contacts List

As I discuss a bit later in this chapter, you can use Alexa to make voice and video calls. You can initiate these tasks by specifying the other person's phone number, but it's almost always easier and faster to specify the other person's name. Alexa is smart, sure, but it's not smart enough to know the name of your best friend or your sister. To give Alexa those smarts, you need to give it access to your contacts. Note, too, that Alexa can send text messages only to other Alexa users, and for that to happen Alexa needs to examine your contacts to see which ones use Alexa.

Giving Alexa access to your contacts

How you give Alexa permission to use your contacts depends on whether you're using the Alexa app on an iOS (or iPadOS) device or an Android device:

>> **iOS (or iPadOS):** Choose Settings ⇨ Amazon Alexa and then tap the Contacts switch to on.

>> **Android:** Choose Settings ⇨ Apps ⇨ Amazon Alexa ⇨ Permissions and then tap the Contacts switch to on.

After you've given the Alexa app permission to access your contacts, you can view those contacts by tapping the Communicate icon and then tapping the contacts icon in the upper-right corner of the Communication screen, labeled in Figure 5-1, left. The Alexa app displays the Contacts screen and shows your profile at the top, as shown in Figure 5-1, right.

Contacts · Manage contacts

FIGURE 5-1:
Tap the contacts icon (left) to view and manage your contacts (right).

It's important to note that Alexa displays your contacts in two ways according to the following two categories:

>> **Alexa-to-Alexa:** Contacts who, like you, also use Alexa's Calling & Messaging feature, which means you can communicate with one of these contacts by sending a text or making a call from your Alexa device to your friend's Alexa device. The Alexa app identifies Alexa-to-Alexa contacts by adding the Alexa Calling & Messaging icons to the contact's screen, as shown in Figure 5-2.

>> **Non-Alexa:** Contacts who don't use Alexa's Calling & Messaging feature, which means you can communicate with one of these contacts by making a call from your Alexa device to your friend's mobile device. Note that you can't send text messages to non-Alexa contacts.

TIP

For easy access to someone you contact frequently, you can add that person to the Favorites list, which appears just below the Add New command (refer to Figure 5-1). Tap the contact you want to work with, and then tap Add to Favorites.

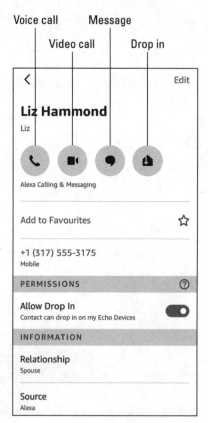

Voice call   Message

Video call   Drop in

<   Edit

Liz Hammond

Liz

Alexa Calling & Messaging

Add to Favourites ☆

+1 (317) 555-3175
Mobile

PERMISSIONS ⑦

Allow Drop In
Contact can drop in on my Echo Devices

INFORMATION

Relationship
Spouse

Source
Alexa

FIGURE 5-2:
An Alexa-to-Alexa
contact has these
Alexa Calling &
Messaging icons.

Inviting a contact to connect via Alexa

If you have contacts who don't use Alexa's Calling & Messaging feature, but you'd like them to at least consider it, you can send them a text message inviting them to use Alexa. Here are the steps to follow:

1. **In the Alexa app, display the Contacts list.**

2. **Tap the contact you want to invite.**

3. **Tap Invite *Contact* to connect on Alexa (where *Contact* is the first name of the contact).**

   The Alexa app creates a new text message that includes the invitation.

4. **Edit the message as you see fit, and then send it.**

   The Alexa app sends the message to your contact.

Giving a contact a nickname

When Alexa has access to your contacts, you normally specify a text or call recipient by including that person's full name in your voice command. That's fine for people you communicate with only occasionally, but it feels a bit formal and a tad long-winded for folks you're in touch with frequently.

You can make it easier and faster to start a conversation with your regular contacts by giving each one a unique nickname that you can use in your calling and texting voice commands instead of the person's full name. Here are the steps to follow:

1. **In the Alexa app, display the Contacts list.**

2. **Tap the contact you want to work with.**

3. **Tap the Add Nickname link, which appears just under the contact's full name.**

   The Edit Contact screen appears.

4. **Use the Nickname text box to type the nickname you want to use for the contact.**

5. **Tap Save to preserve your changes.**

Adding a contact

Alexa communications are easiest when you give it permission to use your mobile device contacts. However, if there's a person who's not in your mobile device contacts (and for some reason you don't want that person there), you can add that person to your Alexa Contacts list directly. Here's how:

1. **In the Alexa app, display the Contacts list.**

2. **Tap Add New ⇨ Add Contact.**

   The Add Contact screen appears.

3. **Enter the person's first name, surname, nickname, relationship, and mobile phone number.**

4. **Tap Save.**

   The Alexa app adds the person to the Contacts list.

Editing a contact

For anyone in the Alexa app's Contacts list, you can tap that person's entry and then tap Edit to open the Edit Contact screen. Besides entering a nickname for that person, as I describe earlier, what else can you do in that screen? It depends:

>> If you added the contact using the Alexa app, you can edit all the information you entered.

>> If the contact was imported from your mobile device's contacts, you can modify most of the information by using the Contacts app on your mobile device. You should see the updated information in the Alexa app's Contacts list within a few seconds. (If the contact you edited is currently open in the Alexa app, close the contact card and then reopen it to see the changes.)

>> If you want to specify your relationship with that person, tap Relationship, tap the relationship you have with that person (such as Spouse, Sister, or Parent), tap Done, and then tap Save.

Blocking a contact

If you have an Alexa-to-Alexa contact that has become tiresome, annoying, or even abusive, you can prevent that pest from contacting your through your Alexa device by using Alexa's blocking feature. Here's how it's done:

1. **In the Alexa app, display the Contacts list.**

2. **Tap the contact you want to block.**

3. **Scroll to the bottom of the contact's card and tap Block Contact.**

   The Alexa app asks you to confirm the block.

4. **Tap Block to confirm.**

TIP

If you have a lengthy Contacts list, there's a quicker way to block someone. In the Contacts screen, tap the manage contacts icon (the three dots in the upper-right corner; refer to Figure 5-1), tap Block Contacts, and then tap Block by Contact Name. You see a list of all your contacts, with a Block link to the right of each person's name. Tap the Block link for the person you want to shun, and then tap Block when Alexa asks you to confirm.

Managing your profile and settings

Alexa's Contacts list also includes an entry for you, which Alexa refers to as your *Calling & Messaging profile*. This entry appears at the top of the list, and tapping it enables you to make a couple of useful changes:

>> **Your name:** To change the name associated with your profile, tap Edit, use the Edit Contact screen to change the First Name and Last Name fields as needed, and then tap Save.

>> **Caller ID:** By default, Alexa says or shows your profile phone number as a caller ID when you make a call. If you'd rather not have your phone number spoken or displayed, tap the Show Caller ID switch to off.

Exchanging Text Messages

If your Contacts list includes people who also use Alexa Calling & Messaging, you're in luck: You can pester those people with Alexa-to-Alexa text messages.

REMEMBER

Note that your text messages are sent through Alexa, not your mobile account, so any messages you ship out are not counted against your mobile account's texting quota.

To do the messaging thing, you can use either voice commands or the Alexa app.

Sending a text using voice commands

Unlike most Alexa interactions, sending a text message to another Alexa user requires several voice commands. It's the closest you can come to having an actual conversation with your virtual personal assistant.

To get started, use any of the following voice commands:

>> "Alexa, send a text message."

>> "Alexa, send a message."

>> "Alexa, send a text."

Alexa responds with, "To whom?"

Now say the contact's name or nickname. Alexa responds with, "Did you mean *name*?" where *name* is the contact Alexa thinks you meant. If that's not the person you want, say, "No," and try again; otherwise, say, "Yes."

Now Alexa prompts you with "What's the message?" Go ahead and recite your message. When you're done, Alexa says, "Got it. Should I send it?" Say, "Yes," and your message is on its way.

Sending a text using the Alexa app

Using voice commands to send a text requires a long conversation with Alexa, so the hands-free texting route is suitable for only short messages. For longer messages, or if you want to give your message a bit more thought as you compose it, you can send the text the old-fashioned way: by typing it.

Typing and sending text requires the Alexa app and the following steps:

1. **In the Alexa app, display the Contacts list.**

2. **Tap the contact you want to message.**

3. **Tap the message icon (labeled in Figure 5-2).**

   If you don't see the message icon, it means the contact isn't using Alexa Calling & Messaging, so I'm afraid you're out of luck.

   The Alexa app opens a conversation screen for the recipient.

4. **In the Type Your Message text box at the bottom of the screen (see Figure 5-3), tap out your message.**

   If you want to record a message instead, tap the microphone icon, say your message, and then tap the Stop button that appears.

5. **Tap Return to send the message.**

FIGURE 5-3:
Use the Type Your Message text box to, well, type your message.

Receiving a text message

When someone sends you a text, Alexa has no shortage of ways to let you know:

>> You hear a tone on your Alexa device.

>> On an Alexa device without a screen, the light ring pulses yellow.

>> On an Alexa device with a screen, you see "Message for *you* from *name*," where *you* is your first name and *name* is the sender's name.

>> On the mobile device that has the Alexa app installed, you see a device notification.

To get your message using your Alexa device, say the following voice command:

"Alexa, play my messages."

To see your message using the Alexa app, tap the Communication tab, which shows the most recent messages you're received from each sender you're talking to. To view the entire conversation, tap a message. In the resulting conversation screen, you might see two types of messages:

>> Standard-issue text messages for messages sent your way via typing them in the Alexa app

>> Voice messages that show both the message text and a Play Message icon (see Figure 5-4) that you can tap to hear the original message

FIGURE 5-4:
When you receive a voice message, tap the Play Message icon to hear the original recording.

12:24 PM

▶ Play Message
Hey Karen are you available for lunch

Type your message...

Making Voice or Video Calls

Alexa devices come with both a speaker and a microphone. What does that remind you of? That's right: a telephone. So it seems fitting that your Alexa device supports a feature called Alexa Calling, which lets you use Alexa to make phone calls.

For example, with Alexa-to-Alexa Calling, you can use your Alexa device to make phone calls to other Alexa devices. Now here's the surprising part: You can also use your Alexa device to make phone calls to any mobile phone (Alexa-to-Mobile Calling) or landline (Alexa-to-Landline Calling) in the United States, Canada, and Mexico. Sweet!

But wait, it gets even better. If you and the person you're calling have Alexa devices with screens, you can also make video calls between those Alexa devices. Sweet with a cherry on top!

What's that? You don't have an Alexa device? Don't sweat it: You can also make Alexa-to-Alexa, Alexa-to-Mobile, and Alexa-to-Landline calls using the Alexa app.

What's the charge? Exactly nothing. What's the catch? There isn't one, although Alexa Calling does have a few restrictions. Specifically, you can't use Alexa to call the following:

>> Emergency services, such as 911

>> All other service codes (also called *N11* numbers), such as 311 and 411

>> Premium-rate numbers, such as toll numbers and 1-900 numbers

>> Dial-by-letter numbers (also called *phonewords*), such as 1-800-FLOWERS and 1-800-GOT-JUNK, although you can instead dial the numerical equivalent of these numbers

>> Non–North American numbers (that is, numbers outside the United States, Canada, and Mexico)

Other than that, you're free to call Aunt Sally in Saskatoon or your cousin Vinny in Vegas. The next few sections take you through the details.

Making voice calls with voice commands

If you have an Alexa device, you can use voice commands to make voice calls. The specific commands to use depend on who (or what) you're calling, so let me take you through the possibilities.

Calling an Alexa device

To place a voice call to another Alexa device, you can use either of the following voice commands:

>> "Alexa, call *name*."

>> "Alexa, call *device*."

Here are some notes about these commands:

>> You can replace the word *call* with *phone, telephone, dial,* or *ring.*

>> For *name,* specify either the person's name or nickname if that person is in your Contacts list and is using Alexa Calling & Messaging.

>> For *device,* specify the name of another Alexa device on your network.

Calling a contact

If the person you want to call is in your Contacts list, you can use any of the following voice commands to call that person's mobile or landline:

>> "Alexa, call *name*'s *label*."

>> "Alexa, call *name* at *label*."

>> "Alexa, call *name* on his/her/their *label* phone."

Here are a few notes to remember when you use these commands:

>> You can replace the word *call* with *phone, telephone, dial,* or *ring.*

>> For *name,* specify either the person's name or nickname.

>> For *label,* specify the label associated with the phone number you want to call. These are labels you assigned to the numbers using your mobile device's Contacts app. Examples include *mobile, home,* and *work.*

Calling a phone number

If you know the phone number you want to call, you can use the following voice command to have Alexa dial the number for you:

"Alexa, call *phone number*."

Replace *phone number* with the numbers you want Alexa to dial, including the area code. Here's an example:

"Alexa, call 3 1 7 5 5 5 1 2 3 4."

If you want, you can replace the word *call* with *phone, telephone, dial,* or *ring.*

Making voice calls using the Alexa app

To make a call using your smartphone or tablet, use the Alexa app to initiate the call. You can go about this in several ways, as you discover in the next three sections.

Calling an Alexa device

If the person you want to speak with is in your Contacts list and is using Alexa Calling & Messaging, you can call that person's Alexa device by following these steps:

1. In the Alexa app, display the Contacts screen.

2. Tap the Alexa user you want to call.

3. Tap the voice call icon (labeled in Figure 5-2).

   If you don't see the voice call icon, the contact isn't using Alexa Calling & Messaging, so you can't call this person's Alexa device.

   The Alexa app places the call to the contact's Alexa device.

Calling a contact

If the person you want to talk to is in your Contacts list, you can call that person by following these steps:

1. In the Alexa app, display the Contacts screen.

2. Tap the contact you want to call.

3. Tap the phone number you want to call.

   The Alexa app places the call to the contact phone number.

Making calls with voice commands

You can use the Alexa app also to initiate phone calls by using voice commands. In the app, tap the Alexa icon and then use any of the following commands, which I discuss in more detail in the "Making voice calls with voice commands" section, earlier in this chapter:

>> "Alexa, call *name*."

>> "Alexa, call *device*."

>> "Alexa, call *name*'s *label*."

>> "Alexa, call *name* at *label*."

>> "Alexa, call *name* on his/her/their *label* phone."

>> "Alexa, call *phone number*."

Making video calls with voice commands

If both you and the person you want to call have either an Alexa device with a screen and a camera or a mobile device that has both a front camera and the Alexa app installed, you can make video calls, where you not only hear each other but also see each other on the device screens. Here are the voice commands you can use:

>> "Alexa, video call *name*."

>> "Alexa, video call *device*."

>> "Alexa, video call *name* on his/her/their *device*."

Some notes about these commands:

>> You can replace the word *call* with *phone, telephone, dial,* or *ring.*

>> For *name,* specify either the person's name or nickname if that person is in your Contacts list and is using Alexa Calling & Messaging.

>> For *device,* specify the name of another Alexa device on your network.

If you see a dialog asking whether Amazon Alexa can use the camera on your device, be sure to tap OK (iOS or iPadOS) or Allow (Android).

Making video calls using the Alexa app

If the person you want to see and hear is in your Contacts list and is using Alexa Calling & Messaging, you can video-call that person's Alexa device by following these steps:

1. In the Alexa app, display the Contacts screen.

2. Tap the contact you want to video-call.

3. Tap the video call icon (labeled in Figure 5-2).

No video call icon in sight? Bummer. It means the contact isn't using Alexa Calling & Messaging, so you can't video-call this person's Alexa device.

The Alexa app places the video call to the contact's Alexa device.

Answering an incoming Alexa-to-Alexa call

If another Alexa user calls you on your Alexa device, you have three ways to answer or ignore the call:

>> **Alexa device without a screen:** When a call comes in, your Alexa device's light ring flashes green, you hear a tone, and Alexa announces "*name* would like to talk" (where *name* is the name of the person calling you). Say, "Answer" to answer the call, or say, "Ignore" if you don't want to talk right now.

>> **Alexa device with a screen:** When a call comes in, your Alexa device plays a tone, you see *name* on the screen, and Alexa announces "*name* would like to talk" (where, in both cases, *name* is the name of the person calling you). To answer the call, either say, "Answer" or tap the green Answer button; if, instead, you don't want to talk, say, "Ignore" or tap the red Ignore button. (The Answer and Ignore buttons you see are the same as the ones labeled in Figure 5-5.)

>> **Alexa app:** When a call comes in, your smartphone or tablet displays its default "Hey, there's an incoming call!" screen, which includes the caller's name. For example, Figure 5-5 shows the screen that appears on an Android phone. To answer the call on an Android device, tap the green answer button; to skip the call, tap the red ignore button; if you're using an iOS (or iPadOS) device, tap Accept to answer the call or Decline to ignore it.

FIGURE 5-5:
An incoming
Alexa call as it
appears on an
Android phone.

Ignore   Answer

Ending a call

If you're talking to someone on your Alexa device, you can use either of the following voice commands to end the call:

>> "Alexa, hang up."

>> "Alexa, end call."

If you're using the Alexa app instead, end the call by tapping the red End button that appears on your mobile device screen (see Figure 5-6).

FIGURE 5-6:
To end a call in the Alexa app, tap the red End button.

End

Dropping In on Other Alexa Users

If you have multiple Alexa devices in your home, it seems like overkill to call someone on another one of those devices. Wouldn't it be easier and faster to establish a direct voice connection with another Alexa device? Why, yes, it would, and that's no doubt why Alexa offers the Drop In feature. With Drop In, you choose a person on your home network that you want to talk to, and Alexa immediately makes the connection.

Wait, I hear you say, isn't that a bit, well, *intrusive*? Yep, it can be. However, people can only drop in on you if you give them permission, so you have some control. The next few sections explain how Drop In works.

Letting folks drop in on you

Before anyone can drop in on your Alexa device, you have to give that person permission. That makes sense because you may not want just anyone dropping in on you. Here are the steps to follow to allow someone to drop in on your Alexa device:

1. **In the Alexa app, display the Contacts screen.**

2. **Tap the contact you want to allow.**

3. **Tap the Allow Drop In switch to on, as shown in Figure 5-7.**

    The Alexa app asks you to confirm.

4. **Tap OK.**

    The Alexa user can now drop in on you.

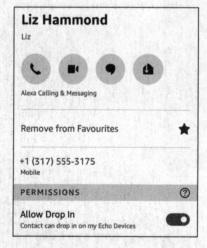

FIGURE 5-7: To let someone drop in on you, open the person's contact card and tap the Allow Drop In switch to on.

Dropping in on someone using voice commands

If an Alexa user has given you permission to drop in, you can make the connection using either of the following voice commands using your Alexa device or the Alexa app:

>> "Alexa, drop in on *name*."

>> "Alexa, drop in on *device*."

Here are some notes about these commands:

>> For *name*, specify either the person's name or nickname if that person is in your Contacts list and is using Alexa Calling & Messaging.

>> For *device*, specify the name of another Alexa device on your network.

Dropping in on someone using the Alexa app

When an Alexa user on your network gives you permission to drop in, that person's contact card in the Alexa app's Contacts list sprouts a new Drop In icon, pointed out in Figure 5-8.

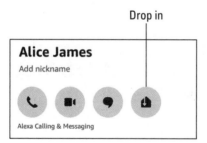

FIGURE 5-8: You see the Drop In icon if a contact has allowed you to drop in.

To drop in on someone using the Alexa app, follow these steps:

1. In the Alexa app, display the Contacts screen.

2. Tap the contact you want to drop in on.

3. Tap the Drop In icon.

If you don't see the Drop In icon, the user hasn't given you Drop In permission or the contact isn't using Alexa Calling & Messaging.

The Alexa app connects to the contact's Alexa device. If you're both using devices with screens, the Drop In connection uses video as well as audio.

Configuring global Drop In settings for your Alexa device

By default, your Alexa device is configured to allow drop-ins only from people to whom you've given Drop In permission, and that applies to every Alexa device on your home network. You can change that to allow only users on your Amazon account or to disable Drop In entirely. Here's how:

1. **In the Alexa app, choose Devices ⇨ Echo & Alexa.**

   You see a list of the Echo and Alexa devices on your Amazon account.

2. **Tap the Alexa device you want to configure.**

   The Alexa device's settings appear.

3. **Tap Communications.**

   The Communications screen appears.

4. **Tap Drop In.**

   The Drop In screen appears, as shown in Figure 5-9.

FIGURE 5-9:
Use the settings in the Drop In screen to configure global Drop In access to your Alexa device.

5. **Select a setting:**

- *On:* The default, this setting allows any Alexa device on your home network, regardless of the Amazon account associated with that device, to drop in, but only if you've given the person using that device permission to do so.

- *My Household:* This setting allows only Alexa devices on your Amazon account to drop in but, again, only if you've given the person using that device permission.

- *Off:* This setting prevents anyone from dropping in on your Alexa device.

TIP

If you want to disable Drop In temporarily, you can put your Alexa device into Do Not Disturb mode. I show you how to do that in Chapter 7.

Using Your Alexa Device as an Intercom

What do you do when the members of your family are scattered around the house and you want to get their attention (say, to call everyone to dinner)? Shout yourself hoarse, right? Well, maybe not. If you have multiple Alexa devices throughout your house, you can make an *announcement*, which is a special Alexa voice message that gets instantly transmitted to every Alexa device on your network. No shouting required (although feel free to do so, if it feels right).

To turn your Alexa device (or the Alexa app) into a one-way intercom, follow these steps in the Alexa app:

1. **Tap the Communicate icon.**

2. **Tap Announce.**

3. **Either type your announcement in the text box provided or tap the microphone icon to record a voice message.**

   If you go the text route, note that you can also add a sound or three to your message by tapping one or more icons in the Add Sounds section.

4. **Tap the announce icon (right-facing arrow, labeled in Figure 5-10).**

   Alexa broadcasts your announcement to each Alexa device on your network.

FIGURE 5-10:
In the Alexa app,
you can type the
message you
want to announce
and add sound
effects.

You can also send announcements using either of the following voice commands:

» "Alexa, announce *message*."

» "Alexa, broadcast *message*."

Change *message* to the announcement you want to make, and Alexa dutifully broadcasts a recording of your voice to every Alexa device on your network. Let's eat!

Asking Alexa to Translate Words and Phrases

Alexa comes with a powerful translation feature that can take a word or phrase in your native tongue and translate it to a language of your choosing, such as Spanish or French:

» "Alexa, how do you say *phrase* in *language*?"

» "Alexa, how do I say *phrase* in *language*?"

» "Alexa, translate *phrase* into *language*."

In each case, replace *phrase* with the word or phrase you want translated, and replace *language* with the target language. As I write this, Alexa can translate into more than 50 languages, with new languages being added regularly.

If you live in the United States, your Echo device can use the Live Translation feature, which enables Alexa to instantly translate between any two languages on-the-fly. To get started, issue the following command:

"Alexa, translate *language*."

Replace *language* with one of the supported Live Translation languages, which as I write these words are the following:

>> Brazilian Portuguese

>> French

>> German

>> Hindi

>> Italian

>> Spanish

When you hear Alexa's beep, start talking. If you talk in English, Alexa translates your words into the language you specified; conversely, if you speak in the specified language, Alexa translates your phrases into English.

When you're finished with Live Translation, say the following:

"Alexa, stop."

Chapter **6**

Using Alexa at Home

lexa is often described as a "virtual personal assistant," and if you haven't felt the truth of that description yet, you almost certainly will by the time you've finished this chapter. That's because over the next few pages, you explore a wide range of Alexa settings, features, and voice commands designed to make your home life easier and more efficient. Do you shop? Do you make a list before you go shopping? Alexa can help with both chores. Do you like to start your day with updates about the latest news, sports, and traffic? Alexa's on it. Do you need help taking care of an aging parent or a loved one with a disability? Just ask Alexa to do these and a bunch of the other tasks that I talk about in this chapter.

A Quick Word about Household Profiles

Before getting to the homey chores, I should mention a couple of Amazon features that, though not related directly to Alexa, can still affect how you and your family use Alexa:

» **Amazon Household:** This feature enables you to share Amazon content (and, for adults, payment methods) with other members of your family. You add family members by using either the Manage Your Household Page (www.amazon.com/myh/manage) or, in selected countries, the Alexa app. For the latter, choose More ➪ Settings ➪ Amazon Household. Either way, you can then add one or more adults, teens, and children to your household. After you've done that, you can switch your Alexa device to another member of the household by saying, "Alexa, switch accounts." (Not sure which household member is using your device? Say, "Alexa, which account is this?" to find out.)

>> **Amazon Kids:** This feature gives you a Parent dashboard that you can use to monitor which content your children are accessing, use parental controls to restrict the types of content your kids can access, and set time restrictions on using Alexa. There's a free version and a paid Amazon Kids+ version that gives you extra controls and content.

Enabling Your Shopping Habit

Amazon's main business is retail, so it's no surprise that the company has created an extensive set of Alexa commands related to buying stuff. You can ask Alexa for recommendations, prices, and the latest deals; you can order new items and reorder previous items; and you can track your orders and get delivery notifications. An old ad for the Yellow Pages used the tagline, "Let your fingers do the walking." Now, with Alexa, you can let your *talking* do the walking.

To use voice shopping with Alexa, you need to set up a few things in advance:

>> The Amazon account you use with Alexa must have a valid payment method and shipping address.

>> Your Amazon account must have 1-Click purchasing enabled.

>> You must be a member of Amazon Prime.

Making sure voice purchasing works

Before you can use voice commands to shop at Amazon, you have to make sure that Alexa's Purchase by Voice setting is turned on. That setting is activated by default, but you should follow these steps to make sure:

1. **In the Alexa app, choose More ⇨ Settings to open the Settings screen.**

2. **Choose Account Settings ⇨ Voice Purchasing.**

3. **Make sure the Purchase by Voice switch is set to on, as shown in Figure 6-1.**

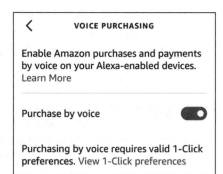

FIGURE 6-1:
The Purchase by Voice switch needs to be on if you want to buy stuff with voice commands.

Getting helpful product suggestions

If you know generally what you're going to buy, but you're not sure exactly what you want, Alexa can help by offering suggestions based on Amazon customer reviews, ratings, and sales. Use any of the following voice commands to get product recommendations from Alexa:

» "Alexa, find the most popular *product*."

» "Alexa, find the best *product*."

» "Alexa, what is the best-selling *product*?"

In each case, replace *product* with the name of the product you want to buy. Alexa responds with a product name and gives you the product's user rating, number of reviews, and price. Alexa then asks, "Would you like to buy it?" Respond with "Yes" to purchase the item.

Ship it: Placing an order

If you know exactly which item you want to purchase, your heart's desire is but a mere voice command away:

» "Alexa, buy *item*."

» "Alexa, order *item*."

» "Alexa, purchase *item*."

Alexa tells you the item it found and then asks, "Do you want to buy it?" Say "Yes" to place the order.

WARNING

If you purchase an item by accident, or just get immediate post-purchase regret, don't get your shirt in a knot: You can usually cancel the item if you act fast. See Chapter 15 for details on how to cancel an Amazon order.

Ship it again: Reordering an item

If you've purchased a particular item in the past, you can make ordering that item again a breeze by using these voice commands:

» "Alexa, reorder *item*."

» "Alexa, order *item* again."

Alexa searches through your order history, suggests an item from that history, and then offers the usual "Do you want to buy it?" prompt. Say "Yes" to reorder the item.

Ship them: Ordering multiple items

If you want to order more than one of a particular item, use one the following voice commands to make it so:

» "Alexa, buy *quantity* of *item*."

» "Alexa, order *quantity* of *item*."

» "Alexa, purchase *quantity* of *item*."

In each case, replace *quantity* with the number you want.

What if you want to order multiple *different* items? Nope, sorry, Alexa doesn't allow that.

Preventing accidental (wink, wink) voice purchases

By default, when Alexa asks, "Do you want to buy it?" and you say, "Yes," your order is placed right away. That can be dangerous if someone else is using Alexa, so you can prevent accidental (or even mischievous) purchases by requiring a four-digit *voice code*.

Wait a second, I hear you saying, if I have to say my voice code for each purchase, anyone nearby (such as my kids) will hear the code, so where's the security in that? Excellent question! If you're worried about a nearby snoop listening for your voice code, you can configure Alexa to authorize purchases only if it recognizes your voice.

Follow these steps to set up your voice code:

1. In the Alexa app, choose More ⇨ Settings to open the Settings screen.

2. Choose Account Settings ⇨ Voice Purchasing.

3. Tap the Voice Code switch to on.

4. Tap Purchase Confirmation.

The Purchase Confirmation screen appears, as shown in Figure 6-2.

5. Select the type of confirmation you want to use for voice purchases:

- *Voice Profile:* Alexa confirms the purchase if it recognizes your voice (or another voice that has a voice profile on the device).

- *Voice Code:* Alexa asks you to say a four-digit voice code to confirm the purchase. When you tap this option, the Alexa app prompts you to enter the code. Tap your four-digit code, and then tap Save.

- *Disable Purchase Confirmation:* Tap this option to make voice purchases without confirmation.

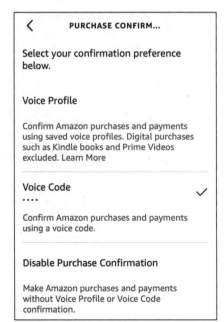

FIGURE 6-2:
Set Recognize Speakers to on to avoid using the voice code when Alexa recognizes your voice.

For the Voice Profile confirmation feature to work, Alexa must have a voice profile for you. If you haven't created a voice profile, select More ⇨ Settings ⇨ Account Settings ⇨ Recognized Voices and follow the on-screen directions.

Did it ship? Checking delivery status

After you've placed an order, you can ask Alexa to give you the current delivery status by using the following voice command:

"Alexa, where's my stuff?"

Alexa responds by telling you the date the shipment is expected to arrive.

If you have multiple orders in the Amazon pipeline, Alexa's vague references to a "shipment" or "package" in the delivery status response can be maddening. Fortunately, you can work around that by configuring Alexa to include item titles in the delivery status response. Here's how:

1. **In the Alexa app, choose More ⇨ Settings to open the Settings screen.**
2. **Choose Notifications ⇨ Amazon Shopping.**
3. **Tap the For Items in Delivery Updates switch to on, as shown in Figure 6-3.**
4. **If you want Alexa to also say the titles of items you've marked as gifts (don't *you* like to live dangerously!), tap the Including Items in Your Shopping Cart Marked as Gifts switch to on.**

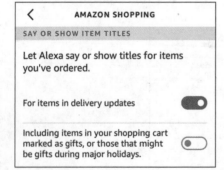

FIGURE 6-3: Tap the For Items in Delivery Updates switch to on to hear item titles.

Yep, it shipped: Getting delivery notifications

Alexa's shopping helpfulness extends to giving you a notification on your Alexa device when your Amazon package has been delivered. Nice. That feature is off by default, however, so follow these steps to turn it on:

1. **In the Alexa app, choose More ⇨ Settings to open the Settings screen.**

2. **Choose Notifications ⇨ Amazon Shopping.**

3. **If you want to get a notification when Amazon releases the shipment for delivery, tap the Out for Delivery switch to on.**

4. **To get a notification when the shipment is delivered, tap the Delivered switch to on, as shown in Figure 6-4.**

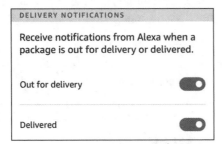

Balancing your karma by making a donation

Voice shopping for yourself or your loved ones makes you feel good, but if you've been a bit indulgent of late, perhaps it's time to make a donation to a favorite charity. I should say make a *voice donation*, because, yep, Alexa is down with donations, too. If you live in the United States, you can use any of the following voice commands to do the donation thing:

- "Alexa, make a donation."
- "Alexa, make a donation to *charity*."
- "Alexa, donate *amount* to *charity*."

Here, *amount* is the amount, in dollars, you want to donate, and *charity* is the name of the charity. If you don't specify an amount or the charity or both in your voice command, Alexa will prompt you for the missing info.

To see a complete list of the charities supported by Alexa, go to `https://pay.amazon.com/us/alexadonations`.

Forging To-Dos and Other Lists

Nod your head if this has happened to you: Your hands are busy or your arms are full, and you suddenly remember something important you need to do. You know that if you don't make a note about the task right away, you'll forget it again. But making a note means either interrupting what you're doing or finding a place to put down your stuff, hunting around for pen and paper or pulling out your smartphone, writing the note, and then resuming. Okay, you can lower your hand now.

Adding items to a to-do list isn't usually that burdensome, but when it is, Alexa can help by offering no-hands-required features that enable you to create and manage lists of things: tasks to do, items to buy, your favorite moments of each day, or whatever.

Creating and managing lists

Alexa helpfully comes with two prefab lists that you can use right away:

>> **Shopping:** A list of items to pick up on your next shopping trip

>> **To-Do:** A list of tasks you need or want to get done

These two lists cover probably 90 percent of list-making for most of us, but it isn't hard to come up with, well, a list of other types of lists. Whatever type of list you need, you can add it to Alexa using either voice commands or the Alexa app.

To create a list with your voice, use either of the following commands:

>> "Alexa, create a list."

>> "Alexa, create a list named *name*."

If you didn't specify the name, Alexa prompts you for it. Alexa then asks for the first item, and you can continue adding items as long as you want.

ALEXA AND THIRD-PARTY LIST APPS

Apps for managing lists are, well, I was going to reach for the "dime a dozen" cliché, but that seems *way* overpriced. Let me just say that *lots* of list apps are out there. However, a few such apps have risen to the top of the, er, list and are worth checking out. Popular list apps include Any.do (www.any.do), AnyList (www.anylist.com), and Todoist (https://todoist.com). Good list managers are also part of *family organizer* apps such as Cozi (www.cozi.com) and Picniic (https://picniic.com).

So, why am I telling you all this in a book about Alexa? Because I bring tidings of great list joy. If you already use — or want to start using — any of the apps I just mentioned, Alexa is cool with that because each app also provides an Alexa skill (see Chapter 9). When you enable a list app skill and then link your list app account with Alexa, any list requests you make of Alexa are automatically applied to your third-party lists.

To make this happen, first sign up for an account with one of the five list apps I mentioned. Then follow these steps to enable the app's skill and link your account to Alexa:

1. **In the Alexa app, choose More ⇨ Settings.**

2. **Tap Lists.**

   The Alexa app displays a screen showing the third-party list apps it supports.

3. **Tap the list app you want to link.**

   The list app's skill page appears.

4. **Tap Enable to Use.**

   The Account Permissions dialog appears. I talk about skill permissions in Chapter 9, but for now all you need to know is that the app is requesting permission to read from and write to your Alexa lists. That's exactly what you want, so leave the two check boxes selected.

5. **Tap Save Permissions.**

   The Alexa app takes you to the list app's login page.

6. **Enter your account credentials for the list app and then log in.**

   The Alexa app links your list app account with Alexa.

To create a list in the Alexa app, follow these steps:

1. **In the Alexa app, choose More ⇨ Lists & Notes.**

   The Alexa app displays the Lists screen. If this is your first time here, you'll see just the default Shopping and To-Do lists.

2. **Tap Create List.**

3. **Use the List Name text box to type the name you want to use.**

4. **Tap Done.**

   The Alexa app adds the new list to the My Lists section of the Lists screen, as shown in Figure 6-5.

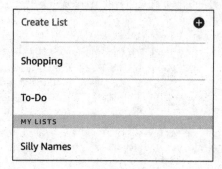

FIGURE 6-5:
Your custom lists appear in the My Lists section of the Lists screen.

You can't rename or delete the Shopping and To-Do lists, but there's lots you can do to mess around with any custom lists you create. In the Alexa app, you can do the following:

TIP

» **View a list:** In the Alexa app's Lists screen, tap a list to see its contents.

   If you have an Alexa device with a screen, you can say, "Alexa, show me my lists," and then tap the list you want to view. Alternatively, say, "Alexa, show me my *name* list" (or "Alexa, what's on my *name* list?") to see the contents of the list named *name*.

» **Rename a list:** In the Alexa app's Lists screen, swipe left on the custom list, tap the Edit button that appears, type the new name, and then tap Done.

» **Share a list:** To email, message, or otherwise share a list with another app or person, open the Lists screen, tap the list, and then tap Share.

» **Print a list:** Navigate to https://alexa.amazon.com, click Lists, click the list you want to print, and then click Print.

» **Archive a list:** If you're done with a list but want to preserve it and its items for some future use, you can archive the list. In the Alexa app's Lists screen, swipe left on the custom list and then tap the Archive button. Alexa archives the list and adds the View Archive link to the Lists screen. Tap View Archive to see your archived lists.

» **Restore an archived list:** To get an archived list back on the main Lists screen, open the Lists screen, tap View Archive, swipe left on the list you want to restore, and then tap the Restore button that appears.

» **Delete a list:** To get rid of a list you no longer need, first archive the list as I describe earlier. Then on the Alexa app's Lists screen, tap View Archive, swipe left on the list you want to remove, and then tap the Delete button that appears.

Adding and managing list items

As I show you in the preceding section, when you create a list, Alexa prompts you to add one or more items to that list. To add subsequent items, use any of the following voice commands:

» "Alexa, add to my *name* list."

» "Alexa, add *item* to my *name* list."

» "Alexa, put *item* on my *name* list."

If you don't specify an item (as in the first command), Alexa prompts you for it.

TIP

If you have an Echo Show, you can add an item to a list also by having the device's camera scan a barcode. Say, "Alexa, scan this to *name* list," hold up the product barcode to the camera, and let Alexa do the rest.

Here are the steps to follow to add an item to a list using the Alexa app:

1. **Choose More ⇨ Lists & Notes.**

   The Alexa app displays the Lists screen.

2. **Tap the list you want to use.**

3. **Tap Add Item.**

4. **Use the Add Item text box to type the item you want to include in the list.**

5. **Tap Done.**

   The Alexa app adds the new item to the list. Figure 6-6 shows the To-Do list with a few items added. Note, too, the Completed section, which displays list items that you've marked as complete, as I describe in the bulleted list that follows.

FIGURE 6-6:
The To-Do list with a few items.

After you have one or more list items, Alexa and the Alexa app offer quite a few techniques for managing your items:

›› **Editing a list item:** In the Alexa app's Lists screen, tap the list and then tap the item you want to change. The item text opens for editing. Make your changes and then tap Done.

›› **Marking an item as complete:** In the Alexa app's Lists screen, tap the list, and then tap the radio button that appears to the left of the item you completed. You can also swipe right on the item or use either of the following voice commands:

- "Alexa, remove *item* from my *name* list."

- "Alexa, check off *item* in my *name* list."

›› **Marking all list items as complete:** Say, "Alexa, clear my *name* list." When Alexa asks you to confirm, say, "Right" or "Yes" or "You go, girl."

›› **Hiding a list's completed items:** In the Alexa app's Lists screen, tap the list, tap More (the three dots in the upper-right corner), and then tap Hide Completed. To restore the completed items, tap the list, tap More, and then tap Show Completed.

›› **Removing a list's completed items:** In the Alexa app's Lists screen, tap the list, tap More, and then tap Clear Completed.

Getting the Information You Need

When you get up in the morning, or perhaps a bit later when you're just about ready to leave the house (and are a bit more awake having had that vital first cup of coffee), you might need to know a few things: the weather (both the current conditions and the forecast), the traffic conditions, the latest news, and whether your favorite team won last night's game.

Oh, sure, you can listen to a local radio station and hope they eventually get around to telling you all that, or you can wrestle with the necessary apps on your smartphone, but as a thoroughly modern citizen of the world, you know there's an easier way, right? Yep, you got it: Alexa was practically *invented* to provide all that information. All you need to know are a few voice commands.

Getting the weather forecast

You can't dress for the weather without knowing what it's like outside, so use the following voice commands to ask Alexa what's going on weather-wise:

>> "Alexa, how's the weather?"

>> "Alexa, what's the weather like?"

>> "Alexa, what's the weather like in *location*?"

>> "Alexa, what's the temperature?"

>> "Alexa, is it raining?"

>> "Alexa, is it snowing?"

>> "Alexa, is it windy?"

For some of these commands (particularly the first two), Alexa goes on to give you the weather forecast. For more specific forecast info, use the following voice commands:

>> "Alexa, what's the extended weather forecast?"

>> "Alexa, will I need an umbrella today?"

>> "Alexa, what will the weather be like *time*?" (For example, "Alexa, what will the weather be like at 6 p.m.?" "Alexa, what will the weather be like this afternoon?" or "Alexa, what will the weather be like tonight?")

>> "Alexa, what will the weather be like *day*?" (For example, "Alexa, what will the weather be like today?" "Alexa, what will the weather be like tomorrow?" "Alexa, what will the weather be like Saturday?" or "Alexa, what will the weather be like this weekend?")

>> "Alexa, will it rain *day*?"

>> "Alexa, will it snow *day*?"

You can also ask Alexa to let you know whenever a severe weather warning has been issued for your city or town:

"Alexa, tell me when there's a severe weather alert."

If you no longer want Alexa to check for these warnings, you can turn off this feature with the following command:

"Alexa, cancel (or turn off) severe weather alerts."

Checking traffic

Alexa can also give you reasonably current traffic information for your commute between home and work. Wait a minute: How does Alexa know where you work? Short answer: It doesn't. That's good news because, well, if Alexa could figure out where you work, that would be downright creepy. No, instead you need to let Alexa know the address of your place of employment. You do that in the Alexa app by following these steps:

1. **Choose More ⇨ Settings.**

2. **Tap Traffic.**

   The Traffic screen appears. You should see your home address as the From value. Alexa assumes you're commuting from home, but that may not be the case if, say, you're traveling or running an errand from another location.

3. **To change the starting point, tap From, enter the new starting address, and then tap Save.**

4. **If you'll be making a stop along the way, tap Add Stop, enter the stop address, and then tap Save.**

5. **To add your ending point, tap To, enter the address of your destination, and then tap Save.**

With your commute info saved, you can now ask Alexa to give you a traffic update. Here are the voice commands you can use:

» "Alexa, how's traffic?"

» "Alexa, what's traffic like?"

» "Alexa, what's my traffic information?"

» "Alexa, what's my commute look like?"

» "Alexa, how's the traffic to *destination*?"

Use that last command if you're driving to somewhere other than work by replacing *destination* with the name or address of where you want to go.

Alexa responds with the approximate time of the journey and some suggested streets to take. Don't want to memorize or write those directions down? Of course not! Instead, say, "Alexa, send that to my phone" and you'll get a notification on your smartphone that, when tapped, opens the directions using your phone's default navigation app.

If you have multiple navigation apps on your phone, you can follow these steps to set your preferred app as the default for Alexa's traffic suggestions:

1. **Choose More ⇨ Settings.**

2. **Tap Traffic.**

3. **Tap Default Navigation App.**

    The Alexa app displays the Navigation App screen, which lists your smartphone's installed navigation apps.

4. **Tap the app you prefer Alexa to use.**

    Alexa uses the app you selected for showing driving directions.

Hearing the latest news with Flash Briefing

If it's news you're after, I recommend Alexa's Flash Briefing feature, which gives you the latest headlines from news sources as well as the most recent daily content offered by certain third-party providers.

The easy part is hearing your Flash Briefing, which you cajole Alexa into playing for you by using the following voice commands:

» "Alexa, play my Flash Briefing."

» "Alexa, next." (Skips to the next item.)

» "Alexa, previous." (Goes back to the previous item.)

The (slightly) harder part is customizing your Flash Briefing to play only the news and updates that you want to hear. Customizing the Flash Briefing means adding extra features to Alexa called *skills*, particularly skills designed to work with the Flash Briefing feature. I go on and on about skills in Chapter 9, but for now you can follow these steps to add content to your Flash Briefing:

1. **In the Alexa app, choose More ⇨ Settings.**

2. **Tap Flash Briefing.**

3. **Tap Add Content.**

   The Alexa app displays a new screen with a list of skills that work with the Flash Briefing.

4. **Tap a skill you think might be a nice addition to your Flash Briefing.**

   The Alexa app displays the skill's info screen, which includes a description of the skill and the skill's user ratings and reviews.

5. **To add the skill to your Flash Briefing, tap Enable to Use.**

   After a moment or two, the Alexa app enables the skill.

6. **Tap the back icon (<) in the upper-left corner to return to the list of skills.**

7. **Repeat Steps 4 through 6 as needed to populate your Flash Briefing.**

   Figure 6-7 shows a Flash Briefing screen with a few skills added.

To manage your Flash Briefing, use the following techniques:

» **Temporarily remove a skill.** On the Flash Briefing screen, tap the skill's switch to off. This moves the skill to the Off section at the bottom of the screen. To restore the skill to your Flash Briefing, scroll down to the Off section and then tap the skill's switch to on again.

» **Reorder the skills.** During the Flash Briefing, Alexa plays the skills in the order in which they appear in the Flash Briefing screen. To get the skills in the order you want to hear them, open the Flash Briefing screen and tap Edit. Tap and hold down on a skill, drag it to the position you prefer, and then release the skill. Repeat as needed and then tap Done.

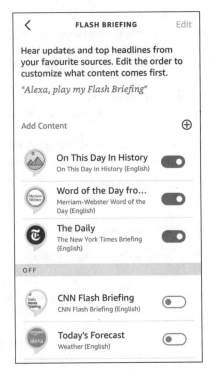

FIGURE 6-7:
A Flash Briefing with some skills added.

Getting long-form news

If you want more in-depth news, Alexa is happy to help. Here's the command to issue:

"Alexa, play news from *source*."

Replace *source* with your favorite news outlet, such as Bloomberg, CBC, CNN, Fox, *The New York Times*, or NPR.

Configuring Alexa's sports update

If you wouldn't even *dream* of going out the door without knowing the latest scores and the upcoming games for your favorite teams, you're going to love Alexa's sports update. To give it a whirl, you need only recite the following command in your best sports announcer voice:

"Alexa, play my sports update."

Alexa responds by giving you each team's most recent result, its next game, and the latest team news. Which teams? Alexa does come with a few default teams, but you'll certainly want to get rid of those and add your own. Here's how to do that:

1. **In the Alexa app, choose More ⇨ Settings.**

2. **Tap Sports.**

3. **To remove a team, tap Edit and then tap the Delete button that appears to the left of the team name. Repeat as needed, and then tap Save.**

   Now you're ready to start adding the teams you follow.

4. **Tap Add a Team.**

   The Alexa app displays the Add a Team screen.

5. **Use the text box provided to start typing the name of the team you want to add.**

   As you type, the Alexa app displays a list of the matching team names.

6. **When you see the team you want to add, tap it.**

   If you want to add multiple matching teams, tap each one to select it.

7. **Tap Save.**

   The Alexa app updates your list of sports teams.

Getting sports news from ESPN

Earlier (see "Getting long-form news") I mentioned that Alexa can provide you with expanded news stories. That long-form goodness applies also to sports news, which you can get via ESPN with the following command:

"Alexa, play news from ESPN."

Setting Up Alexa Care Hub

Do you have a loved one for whom you're the primary caregiver? If so, and if your loved one doesn't live with you, I'm sure you know all too well how difficult it is to provide care when you can't be there. Sure, you can phone or text the person, but (teenagers excepted) no one wants to make (or receive) a large number of phone calls and text messages every day.

Amazon proposes to help solve this remote care conundrum by offering a free service called Alexa Care Hub. When you connect a loved one to Alexa Care Hub, the Care Hub page in the Alexa app shows you each time your loved one interacts with Alexa. You don't see specific activities (such as what music they play or questions they ask), just generic alerts.

Your loved one needs the following for all this to work:

>> An Amazon account

>> An Alexa-enabled device

>> A mobile phone to receive a verification code that Amazon sends as part of the Care Hub setup

Also, Alexa Care Hub is a one-to-one service, meaning you can set it up between only two people: the caregiver and the person being cared for.

REMEMBER

As I write this, Alexa Care Hub is available only in the United States.

To set up Care Hub, follow these steps:

1. **In the Alexa app, select More ⇨ See More ⇨ Care Hub.**

2. **Click Get Started.**

   The Alexa app displays some introductory information about Care Hub. Read through this info until you get to the I Want to Use Care Hub To screen, shown in Figure 6-8.

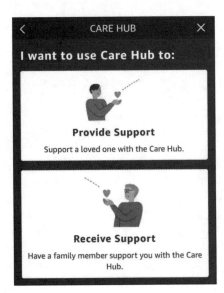

FIGURE 6-8:
You can set up Care Hub for a loved one or for yourself.

3. **Tap a setup option:**

- *Provide Support:* Tap this option if you're the care provider. Use the screen that appears to send a Care Hub invitation to the person you're caring for.

- *Receive Support:* Tap this option if you have someone who cares for you. Use the screen that shows up to send a Care Hub invitation to that person.

  Either way, Alexa sends an email with a Care Hub setup link.

4. **The person who receives the Care Hub invitation email clicks the link and then follows the instructions to complete the setup.**

Once Alexa Care Hub is up and running, you can open the Alexa app and select More ⇨ See More ⇨ Care Hub. If you're the caregiver, you see a list of your loved one's activities so far today; if you're the person being cared for, you see which of your Alexa activities are being shared with your caregiver.

Chapter **7**

Being More Productive

s your life busy? Hah, trick question! *Everybody's* life is busy. Not just busy — *crazy* busy. Our calendars are bursting at their digital seams. Our nagging smart-phones beep and bing at us constantly. Our to-do lists are creating their own to-do lists. Yep, it's a madhouse out there, but that's life in the big city, am I right?

I'm afraid I am. However, that doesn't mean you have to face the onslaught on your own. If you've always wanted a keen and friendly assistant to help you get things done, look no further than your nearby Alexa-enabled device. Sure, Alexa is awesome at solving pop culture arguments and letting you know the latest weather and traffic, but Alexa offers a lot more. On the productivity side, Alexa can help you schedule appointments, let you know when a meeting is getting close, remind you of things you need to do, set alarms, run timers, and more.

In this chapter, you investigate these and other Alexa productivity boosters. Alas, by the end of the chapter, your life will be every bit as busy as it is now, but at least you'll have Alexa sweating some of the small stuff for you.

Letting Alexa Manage Your Calendar

No device can call itself a "personal assistant" without being able to access and manage your appointments, meetings, trysts, special occasions, and other events that populate your calendar. Alexa, I'm happy to report, has earned its "Calendar" badge and can help you manage your schedule so you'll never miss another PTA meeting or recital.

Thankfully, Alexa is smart enough not to offer its own calendar feature — it knows that you almost certainly have your own. In fact, Alexa is ready, willing, and able to work with an existing account from any of the following:

REMEMBER

>> **Apple:** The Calendar component of your iCloud account. Note that Apple requires that your account use two-factor authentication before you can give Alexa access to your iCloud calendar.

Two-factor authentication means that after you enter your account username and password, you have to provide a second account verification, usually a numeric code sent to your phone in a text message.

>> **Google:** The Calendar feature that comes with your Gmail or Google Workspace account. Alexa will also link to your Gmail or Google Workspace email account.

>> **Microsoft:** The Calendar component of your Microsoft 365 subscription. Alexa will also link to your Microsoft 365 email account.

>> **Microsoft Exchange:** The Calendar folder of your Microsoft Exchange server account. To link this calendar, you need to be using Alexa for Business (see https://aws.amazon.com/alexaforbusiness).

You can add more than one calendar type, but Alexa can add events to only a single calendar at a time.

Linking your calendar to Alexa

Before you can use Alexa to manage your schedule, you need to link an existing calendar to Alexa. *Linking* means giving Alexa permission to access and work with your calendar. Here are the steps to follow to link a calendar to Alexa:

1. **In the Alexa app, choose More ➪ Settings.**

2. **Tap Calendar & E-mail.**

   The Alexa app displays the available account types, as shown in Figure 7-1.

3. **Tap an account type.**

   If the account links Alexa to both calendar and email, the Alexa app asks which of these services you want to include.

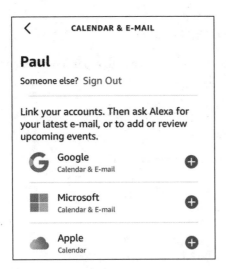

CALENDAR & E-MAIL

FIGURE 7-1:
The account types that Alexa can work with.

4. **If you don't want to include a service, tap its switch to off.**

5. **Tap Next.**

   The calendar provider displays a sign-in screen.

6. **Follow the instructions to sign in to your account.**

   The calendar provider then displays a screen that details what you're giving Alexa permission to do (such as read and update your calendar).

7. **Tap Yes or Allow or whatever button the provider wants you to tap to give Alexa permission.**

Setting the default calendar

If you linked two or more calendars to Alexa, you need to specify the default, which is the calendar to which Alexa will add new events. Here's how to specify the default calendar:

1. **In the Alexa app, choose More ⇨ Settings.**

2. **Tap Calendar & E-mail.**

3. **In the New Calendar Events section, tap the displayed account and then tap the calendar you want to use as the default.**

Unlinking a calendar

If you no longer want to give Alexa access to a particular calendar, follow these steps to unlink that calendar:

1. **In the Alexa app, choose More ⇨ Settings.**
2. **Tap Calendar & E-mail.**
3. **Tap the type account you want to unlink.**
4. **Tap Unlink Account.**

   The Alexa app asks you to confirm.
5. **Tap Unlink.**

   The Alexa app unlinks the account.

Managing calendar events

When you have a calendar linked to Alexa, you're ready to manage your schedule hands-free. The next few sections take you through Alexa's voice commands for managing calendar events.

Adding an event

To create a new event on your calendar, use any of the following voice commands:

>> "Alexa, add an event to my calendar."

>> "Alexa, create a new appointment."

>> "Alexa, add a *duration* event to my calendar."

>> "Alexa, add *event* to my calendar for *date* at *time*."

If you don't specify the event name, date, or time, Alexa will prompt you for the missing specifics. If you don't specify a duration, Alexa defaults to a one-hour event.

Getting upcoming events

A calendar isn't much good if you don't know what events you have coming up, am I right? Alexa will let you know what's on tap if you pester it with any of the following voice commands:

>> "Alexa, what's on my calendar?"

>> "Alexa, what's on the schedule?"

>> "Alexa, show me my calendar" (for Alexa devices with a screen)

>> "Alexa, when is my next event?"

>> "Alexa, what is my next appointment?"

Getting events for a specific time or day

If you're looking to make plans down the road, you need to know what's on your schedule for a certain day or time or both. Use the following voice commands to get your events for a time or day you specify:

>> "Alexa, what's on my calendar *day*?" (For example, "Alexa, what's on my calendar today?" "Alexa, what's on my calendar tomorrow?" "Alexa, what's on my calendar Friday?" "Alexa, what's on my calendar this weekend?" or "Alexa, what's on my calendar December 31st?")

>> "Alexa, what's on my calendar *time*?" (For example, "Alexa, what's on my calendar at 6 p.m.?" "Alexa, what's on my calendar this afternoon?" or "Alexa, what's on my calendar tonight?")

>> "Alexa, what's on my calendar *day time*?"

Moving an event

If your plans change or you entered the wrong time or day for an event, Alexa can move the event to the new or correct time or date or both. Here are the voice commands to wield:

>> "Alexa, move my *event* to *time*."

>> "Alexa, move my *event* to *day*."

>> "Alexa, reschedule my *time* event to *time*."

>> "Alexa, reschedule my *day* event to *day*."

Deleting an event

If an event is cancelled or added inadvertently, or you simply don't feel like going, use any of the following voice commands to clear that event off your schedule:

>> "Alexa, delete my *time* event."

>> "Alexa, cancel my *day* event."

>> "Alexa, delete *event* from my calendar."

Scheduling a meeting with a contact

If you want to set up a meeting, lunch date, or other get-together with someone in your Alexa Contacts list, here are the voice commands to use:

» "Alexa, schedule *event* with *contact*."

» "Alexa, create a meeting named *event* with *contact*."

» "Alexa, invite *contact* to *event*."

Setting Alexa Alerts

Some of Alexa's handiest hands-free features are the alerts you can set. Need to be reminded to tickle your dog at 3 p.m.? Alexa will do that for you (remind you, that is — not tickle the dog). Need to set a daily alarm for 7 a.m. because otherwise you'll *never* get out of bed in the morning? Alexa hears you. Need to set a timer to end your yogic flying session after 30 minutes? Alexa will do that without judgment.

Getting Alexa to remind you of something

If you need to be prodded to perform a task at a certain time (and, optionally, on a particular day), use any of the following voice commands to get Alexa to add a reminder:

» "Alexa, set a reminder."

» "Alexa, set a repeating reminder."

» "Alexa, remind me to *task*."

» "Alexa, set a repeating reminder to *task*."

» "Alexa, remind me to *task* at *time*." (For example, "Alexa, remind me to pet the dog at 7 a.m.," "Alexa, remind me to call Mom at noon," or "Alexa, remind me to eat milk and cookies at 3 o'clock this afternoon.")

» "Alexa, remind me to *task day* at *time*." (For example, "Alexa, remind me to go to the grocery store tomorrow at 3 p.m.," "Alexa, remind me to practice the piccolo Wednesday at 7 p.m.," "Alexa, remind me to mow the lawn next Monday," or "Alexa, remind me to pay my taxes on April 15th.")

If you leave out the task, time, or day, Alexa prompts you for the missing particulars.

To be, uh, reminded of your reminders, just ask Alexa:

» "Alexa, what's my next reminder?"

» "Alexa, what are my reminders?"

You can also set a reminder by using the Alexa app:

1. **Choose More ⇨ Reminders.**

The Alexa app opens the Reminders screen.

2. **Tap Add Reminder.**

The Alexa app displays the Add Reminder screen, shown in Figure 7-2.

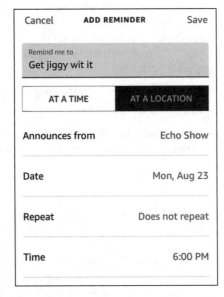

FIGURE 7-2:
Use the Add
Reminder screen
to spell out the
details of your
reminder.

3. **In the Remind Me To text box, enter the task you want to be reminded to perform.**

4. **Tap At a Time.**

5. **If you have multiple Alexa devices, tap Announces From and then select the device you want to use for the reminder.**

TIP

If you have multiple Alexa devices and you want to make sure you don't miss a reminder, you can configure Alexa to announce each reminder on all your devices. In the Reminders screen, tap the gear icon (settings) in the upper-right corner, and then tap the Announce on All Devices switch to on.

6. **If you want Alexa to repeat this reminder on a schedule, tap Repeat to open the Repeat screen, and then tap how often you want the reminder to occur (such as Daily, Weekly, or Monthly).**

Two options in the Repeat screen require some extra steps:

- *Multiple Times a Day:* Tap this option to repeat the reminder two or more times each day. In the Multiple Times screen that appears, tap Repeat and then tap the number of times per day you want to hear the reminder. Alexa suggests a time for each reminder. Edit one or more of the suggested times, if you feel like it, then tap Done.

- *Custom:* Tap this option to create your own schedule. In the Custom screen that appears, use the first Repeat list to select the number of repetitions, and then use the second Repeat list to select the repeat interval (Times a Day, Days, Weeks, or Months). Fiddle with Alexa's suggested times as needed, and then tap Done.

If you're using a repeat interval and you no longer want the reminder to repeat, tap the Does Not Repeat option in the Repeat screen.

7. **If you're not repeating the reminder, use the Date field to specify the date of the reminder and the Time field to specify the time of the reminder.**

8. **Tap Save.**

The Alexa app saves your reminder and Alexa will announce the reminder on the date and time you specified.

TIP

By default, Alexa announces each reminder twice. That second announcement helps make sure you don't miss the reminder, but if it bugs you, you can change the number of times Alexa announces the reminder. In the Reminders screen, tap the gear icon (settings) in the upper-right corner, tap Announcement Count, and then tap the number of times you want Alexa to announce the reminder: 1, 2, 3, or Persistent (that is, Alexa announces the reminder every few seconds for up to an hour).

You can use the Alexa app also to manage your reminders. Tap More, tap Reminders to open the Reminders screen, and then do any the following tasks:

- **»** **Change a reminder.** Tap the reminder to open it in the Edit screen. Make your modifications to the reminder, and then tap Save.

- **»** **Mark a reminder as completed.** Swipe right on the reminder.

- **»** **Delete a reminder.** Swipe left on the reminder and then tap Delete. You can also say, "Alexa, delete my *task* reminder."

TIP

If you want to cancel all your reminders, you can do so in one fell swoop by intoning the following voice command:

"Alexa, delete all my reminders."

Creating location-based reminders

Getting Alexa to prompt you about something at a specified time is darned useful, but Alexa can also remind you of things based on your location. That is, you can set up a reminder to trigger when you either arrive at that location or leave there. When you supply Alexa with a location, it creates a virtual perimeter — called a *geofence* — around that location. When you cross that perimeter, Alexa wakes up and announces the reminder.

For location-based reminders to work, you first need to give the Alexa app permission to always use your location:

- **»** **Android:** Open Settings. Tap App Permissions, tap Location, and then tap the Amazon Alexa switch to on.

- **»** **iOS (or iPadOS):** Open Settings. Tap Privacy, Location Services, Amazon Alexa, and then tap Always. Also, tap the Precise Location switch to on.

With that done, you can use Alexa commands such as the following to set location-based reminders:

- **»** "Alexa, set a reminder for when I leave home."

- **»** "Alexa, set a reminder for when I arrive at work."

- **»** "Alexa, remind me to *task* when I leave my current location."

- **»** "Alexa, remind me to *task* when I arrive at *location*."

You can also follow these steps to create a location-based reminder using the Alexa app:

1. **Choose More ⇨ Reminders.**

   The Alexa app opens the Reminders screen.

2. **Tap Add Reminder.**

   The Alexa app displays the Add Reminder screen (refer to Figure 7-2).

3. **Use the Remind Me To text box to enter the task you want to be reminded to perform.**

4. **Tap At a Location.**

5. **Tap Add Location and then select the location you want to use.**

   In the Add Location screen, you can select Current Location to use your present whereabouts. Otherwise, select your home or work address (go ahead and add those addresses now, if you haven't already). You can also enter a custom address.

6. **Tap When This Phone and then tap either Arrives or Leaves.**

7. **Complete the reminder following the same steps as outlined earlier in the preceding section, "Getting Alexa to remind you of something."**

8. **Tap Save.**

   The Alexa app saves your reminder and Alexa will announce the reminder when you arrive at or leave the selected location.

Seeing notifications for missed reminders

When Alexa announces a reminder, it also sends a notification to your smartphone. This is a welcome feature because it means you still see the reminder even if you're no longer within earshot of your Alexa device. And if you're away from all your devices for a time (good for you!), you can still see your missed reminders by checking your smartphone's recent notifications.

TIP

You can tell you've missed a reminder or two also by checking out your Alexa device: If you see a yellow light, it means reminders were announced. To hear them, say, "Alexa, what are my reminders?"

If you don't like having reminders sent as notifications to your phone, here's how to turn off this feature:

1. **Choose More ⇨ Reminders.**

   The Alexa app opens the Reminders screen.

2. **Tap the gear icon (settings).**

   The gear icon is in the upper-right corner of the Reminders screen.

3. **Tap the Send Notifications to My Phone switch to off.**

   Alexa will no longer send a reminder notification to your smartphone.

Setting an alarm

Whether you want to wake up from a deep sleep or a light nap, use any of the following voice commands to get Alexa to set an alarm:

>> "Alexa, set an alarm."

>> "Alexa, set a repeating alarm."

>> "Alexa, set an alarm for *time*."

>> "Alexa, wake me at *time*."

>> "Alexa, set an alarm at *time* on *device*."

>> "Alexa, set an alarm for *day* at *time*."

>> "Alexa, set a repeating alarm every *interval* at *time*." (For example, "Alexa, set a repeating alarm every day at 8 a.m.," "Alexa, set a repeating alarm every weekday at 6:30 a.m.," "Alexa, set a repeating alarm every weekend at 10 a.m.," or "Alexa, set a repeating alarm every Monday at noon on Echo Show.")

If you leave out the time, the day, or the repeat frequency, Alexa prompts you for the missing info.

Want to wake up to some music instead of the default alarm sound? Sure you do! Here are some commands you can use:

>> "Alexa, wake me up to *song* at *time*."

>> "Alexa, wake me up to *artist* at *time*."

>> "Alexa, wake me up to *genre* at *time*."

>> "Alexa, wake me up to *playlist* at *time*."

You can augment any of these commands with a day, a repeat interval, or a device, as I describe at the beginning of this section.

TIP

If you want to wake up to music from a service other than your default music provider, include the music service in your command. For example, "Alexa, wake me up to *playlist* on Spotify at *time*."

If you can't remember when your alarms are set, ask Alexa:

» "Alexa, what's my next alarm?"

» "Alexa, what are my alarms?"

You can also use the Alexa app to see your set alarms: Choose More⇨Alarms & Timers and then tap the Alarms tab. Tap Add Alarm to open the Add Alarm screen (see Figure 7-3), and then set your alarm time, device, repeat interval, date, and sound. Tap Save when you're done.

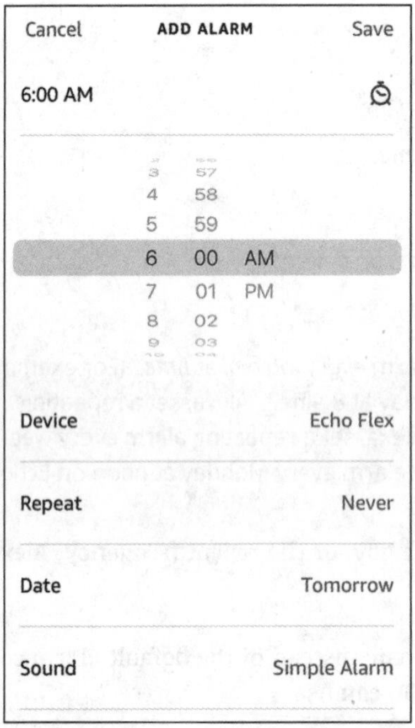

Cancel	**ADD ALARM**	Save
6:00 AM		⏱

3	57	
4	58	
5	59	
6	**00**	**AM**
7	01	PM
8	02	
9	03	

Device	Echo Flex
Repeat	Never
Date	Tomorrow
Sound	Simple Alarm

FIGURE 7-3:
Use the Add Alarm screen to create an alarm.

From the Alarms tab, you can also use the following techniques to manage your alarms:

» **Change an alarm.** Tap the alarm, make your changes, and then tap Save.

» **Disable an alarm.** Tap the alarm's switch to off.

» **Delete an alarm.** Swipe left on the alarm and then tap Delete.

You can get rid of all your alarms at once by using the following voice command:

TIP

"Alexa, delete all my alarms."

When an alarm sounds, you can grab nine more minutes of precious sleep by saying, "Alexa, snooze." To shut off the alarm when it sounds, say, "Alexa, stop" or "Alexa, cancel."

Running a timer

A *timer* — more properly known as a *countdown timer* — is an alert that goes off after a specified duration. Whether you're steeping a cup of tea for 4 minutes or baking a loaf of bread for 40 minutes, use a timer to make sure you don't forget. Alexa does timers with ease and lets you create multiple named timers. Use any of these voice commands to set a timer:

>> "Alexa, set a timer."

>> "Alexa, set a timer for *duration*." (For example, "Alexa, set a time for 4 minutes.")

>> "Alexa, set a timer on *device*." (For example, "Alexa, set a timer on Echo Show.")

>> "Alexa, set a *name* timer." (For example, "Alexa, set a bread timer.")

>> "Alexa, set a *name* timer for *duration*." (For example, "Alexa, set a bread timer for 40 minutes.")

If you leave out the duration, Alexa prompts you for it.

If you can't remember what timers are running, ask Alexa:

>> "Alexa, what are my timers?"

>> "Alexa, how much time is left on the timer?"

>> "Alexa, how much time is left on the *name* timer?"

>> "Alexa, show me the *name* timer" (on a device with a screen).

You can also use the Alexa app to check out your running timers: Choose More ⇨ Alarms & Timers and then tap the Timers tab (see Figure 7-4).

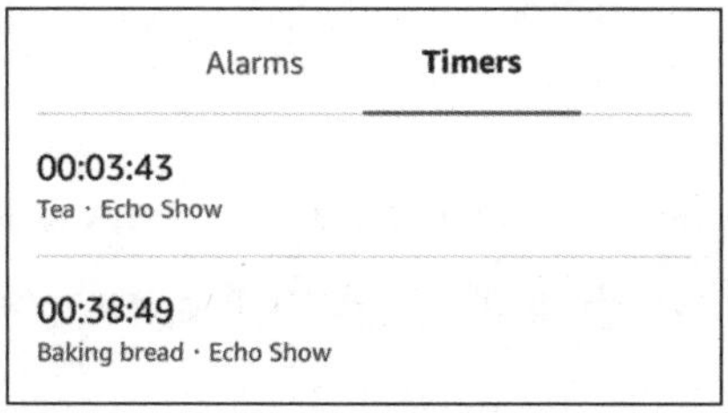

Alarms	**Timers**
00:03:43	
Tea · Echo Show	
00:38:49	
Baking bread · Echo Show	

FIGURE 7-4: In the Alarms & Timers screen, tap the Timers tab to see your running timers.

To delete a timer, you have three choices:

>> If you have a single timer running, say, "Alexa, delete the timer."

>> If you have multiple timers running, say, "Alexa, delete the *name* timer."

>> In the Alexa app, tap the timer and then tap Delete Timer.

TIP

You can cancel all your timers using the following voice command:

"Alexa, delete all my timers."

Customizing your alarm sounds

When Alexa alerts you with an alarm, it plays a simple sound effect that comes on at whatever volume your Echo device is using. That's not a great way to wake up, so you can customize the sounds and the volume. For example, you can select a different sound effect, specify the volume level, and ask Alexa to increase the volume gradually for an easier transition to wakefulness.

Here are the steps to follow to customize your alarm sounds:

1. **In the Alexa app, choose Devices ⇨ Echo & Alexa, and then tap the device you're using for the alarm.**

2. **Tap Sounds.**

The Alexa app displays the device's Sounds screen, as shown in Figure 7-5.

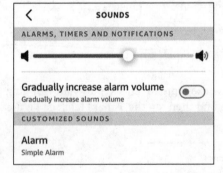

FIGURE 7-5:
The Sounds screen for an Alexa device.

3. **Use the Alarms, Timers and Notifications slider to set the alarm volume.**

Note that this volume level also applies to your timers and to Alexa notifications.

4. **If you want Alexa to slowly raise the volume when the alarm goes off, tap the Gradually Increase Alarm Volume switch to on.**

5. **To change the sound effect that Alexa plays when the alarm goes off, tap Alarm in the Customized Sounds section, and then tap the sound you want to hear.**

   The sound effect plays only if you don't specify music — a song, an artist, a genre, or a playlist — as part of your alarm command.

Putting Alexa in Do Not Disturb Mode

When you combine Alexa's alerts for incoming calls, messages, and drop-ins with the alerts associated with events, reminders, alarms, and timers, you might start thinking that your new virtual assistant is a noisy addition to your home.

You can tone down the noise a bit by putting your Alexa device into Do Not Disturb mode, which silences notifications for incoming voice and video calls, text messages, and drop-ins. Follow these steps to enable Do Not Disturb mode:

1. **In the Alexa app, choose Devices ⇨ Echo & Alexa, and then tap the device you want to put into Do Not Disturb mode.**

2. **Tap Do Not Disturb.**

   The Alexa app displays the device's Do Not Disturb screen.

3. **Tap the Do Not Disturb switch to on.**

4. **(Optional) If you want to control the start and end times for Do Not Disturb mode:**

   a. *Tap the Scheduled switch to on.* The Alexa app adds the Start and End controls, as shown in Figure 7-6.

   b. *Tap Start to set the time when you want Do Not Disturb mode to begin.*

   c. *Tap End to set the time when you want Do Not Disturb mode to finish.*

   Your Alexa device will dutifully go into Do Not Disturb mode at the Start time you specified.

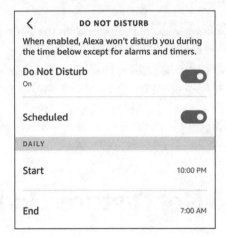

FIGURE 7-6:
You can set up a
schedule for Do
Not Disturb
mode.

Connecting to Someone Else's Echo Device

With the increasing popularity of Amazon's Echo devices, here's a scenario that's becoming more common: You're visiting or staying with a friend or family member who has one or more Echo devices and a situation arises where it would be great to have access to the music, news, or other data on your own Echo device. You're out of luck, right?

Nope. If your friend or family member has enabled a feature called Guest Connect, you can "borrow" that person's Echo device to access your own Alexa data. For this to work, you need the following:

>> An Amazon account

>> An Alexa voice profile

>> A mobile phone number linked to your Alexa profile

>> The Guest Connect feature enabled on the Echo device you want to use

Connecting to another Echo device as a guest

To connect to someone else's Echo device, begin with the following command:

"Alexa, connect my account."

Alexa asks you for your mobile phone number. Say the number, and then wait until you receive a notification on your phone. Open the notification and tap Connect.

When you're done, say the following command to shut off Guest Connect:

"Alexa, disconnect my account."

Enabling Guest Connect on your Echo devices

If you want to allow someone to use Guest Connect on your Echo devices, follow these steps to set it up:

1. **In the Alexa app, select More ⇨ Settings ⇨ Account Settings.**

2. **Tap Guest Connect.**

3. **Tap Allow.**

   Other folks can now use Guest Connect on any of your Echo devices.

3

Getting More out of Your Relationship with Alexa

Pester Alexa with endless questions about nearby places, financial stuff, medical conditions, science topics, math problems, and just about anything under the sun.

Augment Alexa with third-party programs called *skills* that give Alexa new features, new content, new tricks, or new ways to have fun.

Make Alexa accessible for folks who have hearing, mobility, or visual challenges.

IN THIS CHAPTER

» Asking Alexa about nearby places, finances, and medical conditions

» Asking Alexa to solve math and conversion problems

» Asking Alexa about science, culture, and books

» Asking Alexa about dates and times

» Asking Alexa to ask Wikipedia something

Chapter **8**

Asking Alexa Questions

We live in a world where the promise of "information at your fingertips" (IAYF, to those in the know) has gone from pie-in-the-sky daydream just 15 or 20 years ago to let-me-look-that-up-for-you reality now. That's because our fingertips are never far from a device — particularly a smartphone — that's connected to the Internet and its vast supply of information.

But what happens when your fingertips are busy cooking a meal, sewing a dress, sanding a chair, or playing a ukulele? Well, you could wait until you're done. Ha, just kidding, of course. Fortunately, you don't have to wait or do without the information you need right now because you have Alexa right there, by your side. Alexa is on speaking terms (pun intended) with a huge number of facts and figures, and if Alexa doesn't know something, it has a network of sources (such as Wikipedia) that it can dragoon into service. Call it "information at your vocal cords," and in this chapter you unearth the many ways you can ask Alexa what you need to know.

Note that what follows is only the teensiest subset of what you can ask because, technically, you can pretty much ask Alexa about anything and everything. My plan here is to give you a representative sampling of the kinds of questions you can ask. After that, it's up to you and your curiosity.

Searching for Nearby Places

Alexa knows your home address, and that means it can find nearby businesses, such as restaurants and coffee shops as well as an establishment's business address, phone number, and hours. You can find out all this info and more by asking the following questions:

>> "Alexa, what *businesses* are nearby?" (For example, "Alexa, what restaurants are nearby?")

>> "Alexa, what *businesses* are close by?"

>> "Alexa, where's the nearest *business name or type*?" (For example, "Alexa, where's the nearest Starbucks?" or "Alexa, where's the nearest coffee shop?")

>> "Alexa, where can I get *product*?" (For example, "Alexa, where can I get coffee?")

>> "Alexa, what's the address of *business*?"

>> "Alexa, what's the phone number of *business*?"

>> "Alexa, what are the hours of *business*?"

>> "Alexa, is *business* open?"

>> "Alexa, how far is *business*?"

Asking for Financial Info

You probably don't want to get crucial financial information from Alexa, but if you just want some basic info, the following questions can help:

>> "Alexa, what's the *company* stock price?"

>> "Alexa, what's the price of *commodity*?" (For example, "Alexa, what's the price of gold?")

>> "Alexa, what's the current value of *index*?" (For example, "Alexa, what's the current value of the Dow Jones Industrial Average?")

>> "Alexa, what is *amount currency1* in *currency2*?" (For example, "What is 100 U.S. dollars in British pounds?")

>> "Alexa, what is the real interest rate in *country*?"

>> "Alexa, what is the inflation rate in *country*?"

>> "Alexa, what is the gross domestic product of *country*?"

Getting Symptoms and Other Medical Facts

Alexa has access to some top-tier medical sources, including the Centers for Disease Control and Prevention, the National Institutes of Health, and the Mayo Clinic. So, while I'd caution you not to read too much into Alexa's answers regarding symptoms and treatments (you *really* should see a doctor about that), you can at least rest assured that you're not getting medical advice from some dude in a basement. Here are some medical queries you can pose to Alexa:

>> "Alexa, what are the causes of *disease*?"

>> "Alexa, is *disease* contagious?"

>> "Alexa, what are the symptoms of *disease*?"

>> "Alexa, what are the treatments for *disease*?"

WARNING

Amazon records all your Alexa questions and stores them somewhere in its cloud. Consider deleting any medical queries you make through Alexa. See Chapter 15 to learn how to delete your Alexa recordings.

Asking Alexa to Do the Math

Your Android or iOS smartphone has a calculator that does simple math in portrait mode and more advanced math in landscape mode. That's great, but if you need a quick calculation while your hands are otherwise occupied, Alexa's math skills are quite impressive. Here's a sample of the kinds of math questions you can toss Alexa's way:

>> "Alexa, what is *number operator number*?" (For example, "Alexa, what is 2 plus 2?" "Alexa, what is 1,000 minus 383?" "Alexa, what is 52 times 47?" "Alexa, what is 31 divided by 365?")

>> "Alexa, what is *number1* percent of *number2*?" (For example, "Alexa, what is 18 percent of 85.72?")

>> "Alexa, what is *number* squared?"

>> "Alexa, what is *number* cubed?"

>> "Alexa, what is *number1* to the power of *number2*?"

>> "Alexa, what is the square root of *number*?"

- "Alexa, what is the cube root of *number*?"

- "Alexa, what is the *nth* root of *number*?" (For example, "Alexa, what is the 4th root of 256?")

- "Alexa, what is *number* factorial?"

- "Alexa, what is the cosine of *number*?"

- "Alexa, what is the sine of *number*?"

- "Alexa, what is the tangent of *number*?"

- "Alexa, what is the natural logarithm of *number*?"

- "Alexa, what is *number1* modulo *number2*?"

- "Alexa, pick a number between *number1* and *number2*."

- "Alexa, give me a random number between *number1* and *number2*."

- "Alexa, is *number* a prime number?"

- "Alexa, what's the value of *e*?"

- "Alexa, what's the value of Pi to *number* decimal places?"

- "Alexa, what's the value of Pi?" (**Warning:** Expect a *very* long answer!)

Converting Units

If you want Alexa to convert from one unit to another, you need to know just two general voice commands:

- "Alexa, how many *units1* are in *number units2*?"

- "Alexa, what is *number units2* in *units1*?"

Replace *units1* with the unit measurement you want calculated, and replace *units2* with the unit measurement you know. Here are some examples:

- "Alexa, how many teaspoons are in 1 ounce?"

- "Alexa, how many quarts are in 5 gallons?"

- "Alexa, how many kilograms are in 5.5 pounds?"

» "Alexa, what is 10 degrees Celsius in Fahrenheit?"

» "Alexa, what is 75 degrees in radians?"

» "Alexa, how many U.S. dollars are in 10 British pounds?"

» "Alexa, how many kilometers are in 32 miles?"

Getting Science Answers

If you want to get a quick definition or an overview of a particular science topic, use either of the following general questions:

» "Alexa, what is *topic*?"

» "Alexa, tell me about *topic*."

For more specific queries, you can just ask Alexa anything you want. Here are a few examples to give you the flavor:

» "Alexa, how thick is the Earth's crust?"

» "Alexa, what's the tallest tree?"

» "Alexa, what's the fastest animal?"

» "Alexa, what is the Earth's atmosphere composed of?"

» "Alexa, who won the 2020 Nobel Prize for physics?"

» "Alexa, what's the chemical symbol of Mendelevium?"

» "Alexa, who invented the television?"

» "Alexa, when was the telegraph invented?"

» "Alexa, what's the closest galaxy?"

» "Alexa, how old is the Earth?"

» "Alexa, what is the melting point of lead?"

» "Alexa, how many bones are in the human hand?"

Quizzing Alexa about the Culture

For general culture questions, ask Alexa about the topic, like so:

>> "Alexa, what is *topic*?"

>> "Alexa, tell me about *topic*."

Specific questions you can ask are as varied as the culture itself, which means anything goes. Some examples:

>> "Alexa, what's the birthday roundup?"

>> "Alexa, who is Amal Clooney married to?"

>> "Alexa, when are the Oscars?"

>> "Alexa, who is the prime minister of Canada?"

>> "Alexa, how old is James Corden?"

>> "Alexa, who designed the Chrysler Building?"

>> "Alexa, how many people live in Truth or Consequences, New Mexico?"

>> "Alexa, is chocolate good for me?"

>> "Alexa, when was the War of 1812?"

>> "Alexa, who painted the *Mona Lisa?*"

>> "Alexa, when was Galileo born?"

Bugging Alexa about Books

For a virtual AI who lives in a cloud, Alexa knows a *lot* about books. Whether you want to know more about a favorite author or need to settle a bet, here are a few book-related questions you can shout Alexa's way:

>> "Alexa, who wrote *book*?"

>> "Alexa, when was *book* published?"

>> "Alexa, what is *book*?"

>> "Alexa, what is *book* about?"

>> "Alexa, tell me about *author*."

- "Alexa, how many books has *author* written?"

- "Alexa, what is *author*'s latest book?"

- "Alexa, what is *author*'s next book?"

- "Alexa, what genre is *book*?"

- "Alexa, give me a quotation from *book*."

- "Alexa, who are the main characters in *book*?"

- "Alexa, tell me about *character*." (For example, "Alexa, tell me about Holden Caulfield.")

- "Alexa, who won the *year* Nobel Prize for Literature?"

Asking Linguistic Questions

Alexa has a few useful linguistic tricks up its digital sleeve. Here are some questions you can ask:

- "Alexa, what's the definition of *word*?"

- "Alexa, how do you spell *word*?"

- "Alexa, how do you pronounce *word*?"

- "Alexa, what's a synonym for *word*?"

- "Alexa, what's the opposite of *word*?"

- "Alexa, what rhymes with *word*?"

- "Alexa, what's the longest word in the English language?"

Alexa is also pretty decent at translating English words and phrases and can work with more than 50 languages. Here are some translation-related questions you can pose:

- "Alexa, what languages can you translate?"

- "Alexa, how do you say *word* in *language*?"

- "Alexa, translate *word* in *language*."

- "Alexa, what's the word for *word* in *language*?"

Note that in each case you can substitute a short phrase for the *word* placeholder.

Asking Questions about Companies

If you're looking for some basic facts about a company, Alexa can help. Here are a few general questions you can lob at Alexa:

- » "Alexa, what is *company*?"

- » "Alexa, when was *company* founded?"

- » "Alexa, who founded *company*?"

- » "Alexa, who is the *position* of *company*?" (For example, "Alexa, who is the CEO of Apple?")

- » "Alexa, what is the location of *company*'s headquarters?"

- » "Alexa, what is the stock price of *company*?"

- » "Alexa, what is the market capitalization of *company*?"

Asking about Dates and Times

I never know when Easter falls this year or when daylight saving time begins or ends. Now I don't need to know because I can just ask Alexa. Here are some questions you can ask related to dates and times:

- » "Alexa, what day of the week is *date*?"

- » "Alexa, how many days until *date*?"

- » "Alexa, how many days until *holiday*?"

- » "Alexa, how many days has it been since *date*?"

- » "Alexa, how many years has it been since *year*?"

- » "Alexa, when is *holiday*?"

- » "Alexa, when is the first day of *season*?"

- » "Alexa, when is the *solstice* or *equinox*?"

- » "Alexa, when does daylight saving time start/end?"

- » "Alexa, when is the next full moon?"

For Everything Else: Querying Wikipedia

If Alexa can't find something or doesn't understand a query, there's a good chance that there's an article on Wikipedia on the same topic. So, to close this chapter on asking Alexa stuff, here are the voice commands to use to ask Alexa to return the Wikipedia article on a particular topic:

» "Alexa, ask Wikipedia about *topic*."

» "Alexa, Wikipedia *topic*." (For example, "Alexa, Wikipedia rocket science.")

Chapter **9**

Adding Skills to Alexa

The Alexa you interact with on your Echo or other Alexa-enhanced device, or on your mobile device via the Alexa app, is jaw-droppingly good technology. Sure, once in a while Alexa misinterprets one of your voice commands or tells you that "I can't find X" when you know for a *fact* that X exists, but these are mere quibbles. Even if you're skeptical about artificial intelligence in general, and wary of inviting a "listening device" into your home in particular, you have to admit that Alexa is awfully impressive.

But that impressiveness is just out-of-the-box Alexa. Amazingly, you can make Alexa even better by bolting on extra features. These *skills*, as Amazon calls them, enhance Alexa with new capabilities, such as playing games, teaching you new things, and controlling home-automation products.

In this chapter, you explore the world of Alexa skills by learning what they are, where to find them, and how to enable them on your Alexa device. You even learn how to build your own skills!

Understanding What a Skill Is

When you take your new smartphone or tablet out of its box and turn it on for the first time, the Home screen is plastered with a set of default apps that enable you to perform basic tasks such as managing your contacts and calendar, exchanging email and texts, and browsing the web. If you want to perform more specialized tasks such as hailing a Lyft, finding out what's on Netflix tonight, or playing Words with Friends, you need to open your device's app store and install the appropriate app.

The Alexa equivalent of mobile-device apps are called *skills*. Like a mobile device, Alexa can do quite a few useful tasks by default, such as playing music or audiobooks, telling you the time or the weather, and making voice or video calls. But if you want Alexa to perform more specialized tasks, you need to add the appropriate skills.

What kinds of skills can you add? Well, more than 100,000 skills are available in categories such as Business & Finance, Education & Reference, Games, Trivia & Accessories, and Smart Home, so the sky is pretty much the limit. Want to hail an Uber or a Lyft? There are skills for that. Want to know what's on TV tonight? There's a skill for that. Want to play an adventure game or challenge you and your family with some trivia? There are skills for those, too.

Checking Out Amazon's Alexa Skills Store

Okay, skills certainly sound intriguing, but how do you get them? The skill equivalent of an app store is Amazon's Alexa Skills Store, which you can access in one of two ways:

>> **On the web:** Use your favorite web browser to surf to www.amazon.com (or your local Amazon domain, such as www.amazon.ca for Canada), click the navigation menu (the three-line "hamburger" menu named All) and then choose Echo & Alexa ⇨ Alexa Skills. Figure 9-1 shows a portion of the Alexa Skills page that appears.

>> **In the Alexa app:** Choose More ⇨ Skills & Games. You see a screen similar to the one shown in Figure 9-2.

TIP

By default, the Alexa app doesn't show skills that were designed specifically for children. To see those skills, you need to adjust your settings. Choose More ⇨ Settings ⇨ Account Settings ⇨ Kid Skills, and then tap the Allow Kid Skills switch to on. (If you don't see the Kid Skills item, it means that feature hasn't been rolled out in your location yet.)

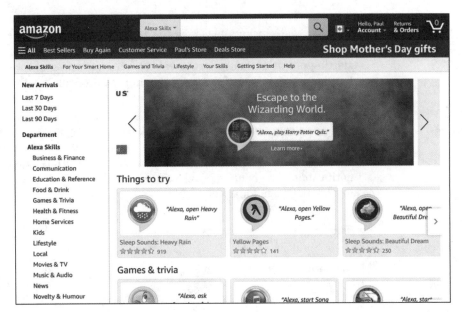

FIGURE 9-1:
Amazon's Alexa
Skills Store.

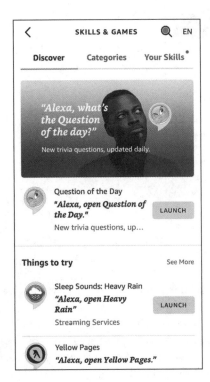

FIGURE 9-2:
The Alexa app's
Alexa Skills
screen.

Whether you're using the web or the app, you see a few featured skills up front, and you also get a search box to locate specific skills. However, when you're just starting out, it's usually best to examine the various skill categories, which are listed on the left side of the web page. In the Alexa app, tap the Categories tab to see the list.

Select a category to see what's available. If something looks interesting, select the skill to open its info page, which gives you a description of the skill, voice commands you can use, user ratings, and any required permissions (such as allowing notifications or giving your address for a delivery skill). Figure 9-3 shows the information page for a typical skill.

FIGURE 9-3:
The info page
for a typical
Alexa skill.

The info you see may include one or more of the following phrases that may have you scratching your head:

>> **Account Linking Required:** The skill requires access to a third-party user account. For example, if you want to use a ride-sharing service such as Lyft or Uber, both of which offer Alexa skills, you must give the skill permission to access your user account on the service, a process called *account linking*.

>> **Account Linking Available:** The skill can access a third-party user account, but account linking isn't required to use the skill.

>> **In-Skill Purchases Available:** The skill offers extra features or content that you can purchase while using the skill.

In-skill purchases are available only through skills obtained via the U.S. version of the Alexa Skills Store.

REMEMBER

LEARNING MORE ABOUT IN-SKILL PURCHASING

You may be thinking that seeing the notification "In-Skill Purchases Available" is like seeing "batteries not included" in the fine print of a toy's packaging. In other words, getting the item is one thing, but making it do something useful is going to set you back some money. Fortunately, that's not true with skills. All skills are free at the beginning, but a few make money through in-skill purchases. However, these purchases are optional, which means the skill still offers at least basic functionality without requiring extra features or content.

So, what kinds of purchases can you expect? It depends on the skill, but all in-skill purchases fall into one of the following types:

- **One-time purchase:** Activates, adds, or unlocks a new skill feature or extra skill content. The added feature or content is available to all your Alexa devices that use your Amazon account, and access to that feature or content is permanent. For example, many games offer in-skill purchases for bonus levels or theme packs.

- **Subscription:** Gives you access to new skill content over a specified time, such as a year for an annual subscription. When the subscription period ends, you must renew to continue receiving new content. For example, most media-related skills — news, magazines, podcasts, and so on — offer in-skill purchases for recurring premium content.

- **Consumable purchase:** Adds a resource to the skill that gets depleted as you use the skill. When the resource runs out, you can purchase the resource again to get a fresh supply. For example, consumables are popular in game skills and include resources such as extra lives, in-game currency, and game hints.

Enabling a Skill

If you locate a skill that looks fun, entertaining, or useful, go ahead and give it a try (hey, it's free, remember?). To use a skill with Alexa, you must enable it.

Enabling a skill with a voice command

To enable a skill with your voice, use either of the following commands:

>> "Alexa, enable *skill*."

>> "Alexa, open *skill*."

Here, *skill* is the so-called *invocation name* of the skill: a word or phrase unique to the skill. How in the name of Jeff Bezos are you supposed to know a skill's invocation name? It's available on the skill's info web page, in the Skill Details section. For example, Figure 9-4 shows the Skill Details section for the popular skill Headspace: Guided Meditation for Everybody, and you can see that its invocation name is *headspace*.

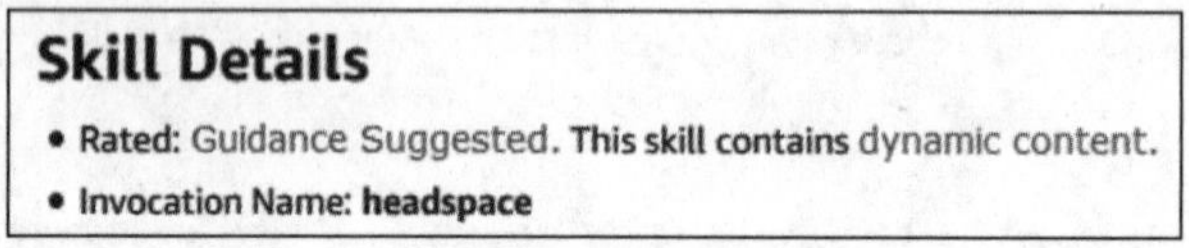

FIGURE 9-4:
See the Skill Details section to get a skill's invocation name.

Skill Details

- **Rated:** Guidance Suggested. **This skill contains** dynamic content.
- **Invocation Name: headspace**

Alexa tells you the skill is enabled and then invokes the skill. The first time the skill runs, you usually get a brief introduction and some suggested commands or tasks.

If the skill requires permission from you to access information on your Alexa device, you can't enable the skill with a voice command. Instead, you have to use the Alexa app, as I describe next.

Enabling a skill using the Alexa app

You can also enable a skill using the Alexa app, as shown in the following steps:

1. **In the Alexa app, choose More ⇨ Skills & Games.**

2. **Locate the skill you want to enable.**

3. **Tap the skill to open its information page.**

4. **Tap Enable to Use.**

   If the skill requires extra permissions from you, you see the Account Permissions dialog. For example, Figure 9-5 shows an Account Permissions dialog for a skill that requires the address information associated with the Alexa device. Another common permission that skills ask for is whether they can send you notifications.

5. **If you see the Account Permissions dialog, select or deselect the permission check boxes, as desired, and then tap Save Permissions.**

   Alexa enables the skill.

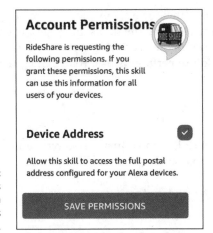

FIGURE 9-5: Some skills require extra permissions from you.

TECHNICAL STUFF

Amazon has announced that it wants to eventually do away with enabling skills. That doesn't mean it's getting rid of skills, however — not even close. Instead, Amazon's goal is to hide skills and have them available in the background, no enabling required. So, how will Alexa know which skill you want? That's hazy at the moment, but the general idea is that Alexa will either learn what skills you prefer or will choose the most relevant skill depending on your request. This is high-end AI stuff, so don't expect the Alexa Skills Store to go away anytime soon.

Working with Skills

Alexa and the Alexa app give you a fistful of ways to use and maintain your skills. There's nothing strenuous or complex about any of this, so the next few sections just take a quick look at what you can do.

Invoking a skill

When you have a skill enabled, you invoke the skill by using one of the following voice commands:

>> "Alexa, open *skill*."

>> "Alexa, launch *skill*."

>> "Alexa, start *skill*."

Replace *skill* with the skill's invocation name.

Note, too, that most skills have their own commands for operating or configuring the skill. To see these commands, open the skill's info page and examine the suggested voice commands shown in the Try Saying section. One common command is to get Alexa to ask the skill to perform a task. Here's the general form of that voice command:

"Alexa, ask *skill* to *task*."

For example, to get the kid skill named Amazon Storytime to tell you a story, you'd use the following command:

"Alexa, ask Amazon Storytime to read me a story."

Viewing your enabled skills

It's great that Alexa makes it so easy to enable skills, but that ease can also lead to a problem: You may forget which skills you've enabled! Hey, it can happen to anyone, but the Alexa app has your back on this one. Follow these steps to view a list of your enabled skills:

1. **Choose More ⇨ Skills & Games.**

2. **Tap Your Skills.**

   The Alexa app displays the Your Skills screen, which lists the skills you've enabled. Figure 9-6 shows an example.

3. **Use the Your Skills screen tabs to manage the list:**

   • **Enabled:** Displays a complete lists of your enabled skills.

   • **Updated:** Displays a list of skills that have released new versions recently.

- **Attention:** Displays a list of skills that have some sort of problem or extra steps that you need to handle. For example, if a skill uses account linking and you haven't completed that step yet, the skill appears on the Attention tab with the phrase *Account linking required.*

- **Blueprint:** You see this tab only if you create your own skills, as I describe later in the chapter in the "Building Your Own Skill" section. The Blueprint tab lists the skills you've built.

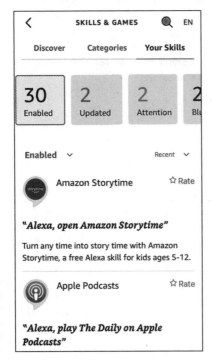

FIGURE 9-6:
The Alexa app's
Your Skills screen
shows you which
skills you've
enabled.

Changing skill settings

Most skills are extremely simple and require only a basic voice command to work. However, some skills have one or more settings you can play around with. Some of the most common settings are activating account linking for skills that require it, managing permissions, and choosing whether you want the skill included in your Flash Briefing, which I described in Chapter 6.

Follow these steps to adjust a skill's settings:

1. **Choose More ⇨ Skills & Games.**

2. **Tap Your Skills.**

3. **Tap the skill you want to work with.**

4. **Tap Settings.**

   The Alexa app displays the Skill Settings screen.

5. **Tap the setting you want to change.**

   In some cases, you tap not the setting name itself but a link that appears beside it. For example, with the Account Linking setting (see Figure 9-7), you tap the Link Account link, which takes you to a page on the skill's website that you use to link your account to Alexa.

6. **Make your changes to the setting.**

7. **If the skill requires permissions, you can tap the Manage Permissions button to adjust those permissions.**

   Alexa saves your new settings.

FIGURE 9-7:
Some settings offer a link that you tap to change the setting.

Disabling a skill

If you find that a skill has become boring, useless, or in some other way a waste of your precious time, you should disable that skill and get it out of your life. If you have your Alexa device nearby, you can use the following voice command:

"Alexa, disable *skill*."

Here, again, *skill* is the skill's invocation name.

WARNING

Alexa doesn't ask for confirmation before it disables the skill, so be sure you have the correct invocation name and that you really do want the skill disabled.

To disable a skill in the Alexa app, follow these steps:

1. **Choose More ➪ Skills & Games.**
2. **Tap Your Skills.**
3. **Tap the skill you want to remove.**
4. **Tap Disable Skill.**

   The Alexa app asks you to confirm.
5. **Tap Disable.**

   Alexa disables the skill.

Thirty Cool Alexa Skills to Try

Yup, I hear you: With more than 100,000 skills available, finding a useful, fun, or entertaining skill is a true needle-in-the-proverbial-haystack exercise. Who has time to comb through either a haystack or Amazon's Alexa Skills Store? If you're just not sure where to begin, let me help. Here are 30 ready-to-enable skills that you can take for a test drive:

» **7-Minute Workout:** Offers daily workouts that aim to improve strength and aerobic fitness in just seven minutes. Invocation name: *seven minute workout.*

» **Akinator:** After asking a few questions, guesses the name of the real or fictional character you're thinking of. Invocation name: *akinator.*

» **Amazon Storytime:** Reads aloud professionally narrated short stories for kids aged 5 through 12. Invocation name: *amazon storytime.*

» **Ambient Sounds:** Plays a collection of sound loops to help you sleep, relax, or block out noises. There are a few dozen sounds in all, including thunderstorms, windy trees, and fireplace. Invocation name: *ambient sounds.*

» **AnyPod:** Enables you to add podcasts to your library, sync your listening history to your Alexa device, and much more. Invocation name: *anypod.*

» **The Bartender:** Delivers cocktail recipes. Invocation name: *the bartender.*

» **Big Sky:** Provides weather info that goes well beyond the default Alexa weather report. Account linking is required. Invocation name: *big sky.*

» **Burglar Deterrent:** Protects your home when you're out by playing realistic audio for activities such as talking, cleaning, and cooking to make it seem as though someone's home. Invocation name: *burglar deterrent.*

» **CBC:** Provides news, music, or a specific Canadian Broadcasting Corporation Radio One station. Invocation name: *CBC.*

» **Curiosity:** Asks you to choose between two topics and then tells you interesting facts about the topic you chose. Invocation name: *curiosity.*

» **The Daily Show:** Delivers a satiric take on the day's news stories, as part of your Flash Briefing. Invocation name: *the daily show.*

» **Headspace:** Provides guided meditations, including a new guided meditation each day. Invocation name: *headspace.*

» **Inspire Me:** Delivers inspirational quotations from famous people (who do the talking instead of Alexa) that you can invoke at random, by speaker, or by topic. Invocation name: *inspire me.*

» **Learn Something Radio:** Plays daily new episodes from some of the web's most learned brands, including NPR, 99% Invisible, and Freakonomics Radio. Invocation name: *learn something radio.*

» **Lyft:** Enables you to use Alexa to hail a Lyft. Account linking is required. Invocation name: *lyft.*

» **The Magic Door:** Plays an interactive adventure game. Invocation name: *the magic door.*

» **Mayo Clinic First Aid:** Provides first-aid instructions for common injuries and illnesses. Invocation name: *mayo first aid.*

» **Meat Thermometer:** Tells you the internal temperature for the type of meat (such as "steak") and how you want it cooked (such as "medium rare"). Invocation name: *meat thermometer.*

- **Mastermind:** Offers an all-purpose AI-driven assistant that you can use to send text messages and emails, make phone calls, search the web, and tons more. Invocation name: *mastermind*.

- **MySomm:** Suggests a wine to go with the food you specify. Invocation name: *wine gal*.

- **NPR News Now:** Announces news stories from National Public Radio, delivered via Alexa's Flash Briefing. Invocation name: *npr news now*.

- **Samuel L. Jackson celebrity voice:** Replaces your regular Alexa voice with the voice of actor Samuel L. Jackson. Invocation name: *Alexa, introduce me to Samuel L. Jackson*.

- **SAT Word of the Day:** Presents a new word each day, spells the word, and uses it in a sample sentence. Invocation name: *s-a-t word of the day*.

- **TED Talks:** Gives you audio access to all the TED Talks. Invocation name: *ted talks*.

- **This Day in History:** Tells you the top historical event that occurred on the current day of the year. Invocation name: *this day in history*.

- **Translated:** Translates short phrases from English into more than three dozen languages, from American Spanish to Welsh. Invocation name: *translated*.

- **True or False:** Gives you a series of questions where the answer to each is either true or false. Invocation name: *true or false*.

- **Uber:** Enables you to use Alexa to hail an Uber. Account linking is required. Invocation name: *uber*.

- **The Wayne Investigation:** Provides an interactive mystery game where the choices you make as you investigate a mystery affect the outcome of the story. Invocation name: *the wayne investigation*.

- **What to Expect:** Delivers daily tips and advice during pregnancy. Invocation name: *what to expect*.

WARNING

Many of these skills are available only in certain countries. If you don't see the skill either in Amazon's Alexa Skills Store or in the Alexa app, the skill isn't available where you live.

Building Your Own Skill

Have you checked out a few skills and thought, "Heck, I can do better than *that!*" Do you — or does someone you know — have a great idea for a skill? Whatever your motivation, building custom Alexa skills is a good news/bad news/good news proposition.

The good news is that, yes, you most certainly can build a custom Alexa skill. The bad news is that building a skill typically requires that you be proficient in a programming language such as Python, C#, or JavaScript. The good news is that you can ignore the bad news! Why? Because Amazon offers a much easier route to a building a custom skill: Alexa Skill Blueprints.

Blueprints are templates that enable you to build a custom skill. More than 70 templates are available in categories such as Fun & Games, Learning & Knowledge, At Home, and Storyteller. Each template takes you step-by-step through the entire process of building your skill. If you can follow a recipe, you can build your own Alexa skill.

You get started by pointing your web browser to `https://blueprints.amazon.com`. Figure 9-8 shows the page that appears. If you see the Sign In button in the upper-right corner, click that button to sign in to your Amazon account.

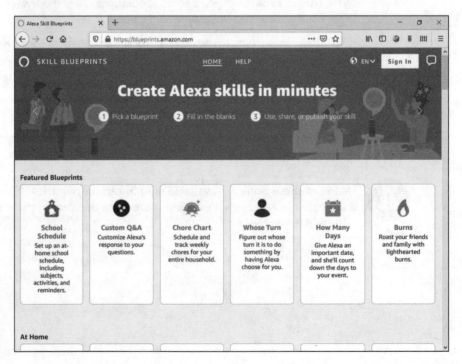

FIGURE 9-8:
Amazon's Alexa Skill Blueprints home page.

Each blueprint has a few unique steps, but a general procedure is common to all blueprints. The following steps take you through that general procedure:

1. **On the Alexa Skill Blueprints home page, click the template for the skill you want to build.**

   You see the template's home page, which offers a short overview, a sample of the skill, and the steps unique to the template. Figure 9-9 shows the home page for the Inspirations template.

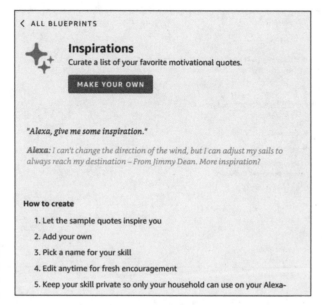

FIGURE 9-9:
The home page
for the Inspira-
tions template.

2. **Click Make Your Own.**

   If you see the Three Steps to a Skill carousel, run through the steps, if you feel like it, and then click Close (X).

3. **For each step in the blueprint, supply the requested information, and then click the Next button.**

4. **When you get to the Name step, edit the template's default name.**

   When entering the name, use two or three relatively simple words that contain only letters, periods, and apostrophes. Don't use your Alexa device wake word or connecting words such as *and, to,* or *for.*

5. **Click Next: Create Skill.**

If this is your first skill, the blueprint tells you that you need to create a free Amazon developer account.

6. **Click Update Account.**

Amazon sets up your developer account and then creates your skill, which takes a few minutes. You'll know your new skill is fully baked when you see the Your Skill Is Created message, as shown in Figure 9-10.

FIGURE 9-10:
A custom Alexa skill, ready for action.

Here are some notes for working with your new skill:

» To invoke your skill, say "Alexa, open *skill*," where *skill* is the name you gave your skill.

» To rename your skill, click the pencil icon that appears to the right of the name, edit the name, and then click Rename Skill.

» To make changes to your skill, click Edit.

» To get rid of the skill, click Delete.

» To share your skill, click Share with Others, click either Yes or No when Amazon asks if the skill is intended for kids under 13 years old, and then click how you want to share the skill: Email, Facebook, Twitter, and so on.

» To publish your skill in the Alexa Skills Store, click Publish to Skills Store and follow the instructions to get your skill ready for prime time and then submitted for review.

176 PART 3 **Getting More out of Your Relationship with Alexa**

Chapter **10**

Making Alexa Accessible

For people with a vision impairment or mobility challenges, Alexa is a gift from the technological gods because devices such as the Echo, Echo Dot, and Echo Flex can be operated by voice only: Good vision or steady hands are not required.

Similarly, for people with severe hearing problems, Alexa is also apparently heaven sent, because devices such as the Echo Show can display Alexa's results on-screen.

Alexa's benefits to people with physical challenges are significant, but they don't mean that either Alexa or an Alexa device is configured ideally for those same people. In this chapter, you explore the accessibility features that can make Alexa easier to use for people with speech, hearing, and vision issues.

Controlling Alexa If You Have a Speech Impairment

Alexa is billed as a voice-activated personal assistant, but that "voice-activated" part is problematic if you can't speak at all or have speech challenges. To fix that, you can take advantage of the many features offered by Alexa and the Alexa app to make Alexa devices accessible to those with speech problems.

Making Alexa more speech accessible

Alexa and the Alexa app come with a few features that you can customize or enable to work around speech impairments:

>> You can use the Alexa app to perform many Alexa tasks by tapping instead of speaking, including creating lists and adding items to lists (see Chapter 6), setting reminders (Chapter 7), viewing contacts (Chapter 5), creating smart-home routines (Chapter 12), and adding skills (Chapter 9).

>> If you have trouble saying "Alexa," you can try a different wake word, as I describe in Chapter 15.

>> If you have an Alexa device with a screen, you can use the Tap to Alexa feature to issue Alexa commands by tapping instead of speaking (see "Using Tap to Alexa," later in this chapter).

>> If you have an Alexa device with a screen, you can swipe left from the right edge of the screen to set alarms, create smart-home routines, and control smart-home devices.

>> If you have an Alexa device with a screen, you can send text messages by tapping instead of speaking (see "Texting without speech," later in this chapter).

Installing a text-to-speech app

One way to work around speech problems with Alexa is to use a text-to-speech (TTS) app on your smartphone or other mobile device. A TTS app lets you type some text, and the app then speaks that text for you through the device speaker. Here are a few popular (and free) TTS apps available for iOS, iPadOS, and Android:

>> Google Translate

>> Talk For Me

>> Text to Speech!

Using Tap to Alexa

TTS apps enable you to interact with screenless Alexa devices such as the Echo, Echo Dot, and Echo Flex. However, if you have an Echo Show, you can use its screen to access a few common Alexa features with just a couple of taps on the screen. This feature is called Tap to Alexa.

When you enable Tap to Alexa, you can tap the Echo Show Home screen to display a grid of icons — Amazon calls them *tiles* — that represent common Alexa tasks: getting the weather or news, setting a timer or alarm, playing music, or adding an item to your Shopping or To-Do list.

To get started, you first have to enable Tap to Alexa by following these steps:

1. **Swipe down from the top of the screen.**

2. **Tap Settings.**

3. **Tap Accessibility.**

4. **Tap the Tap to Alexa switch to on.**

5. **Tap Continue.**

With Tap to Alexa enabled, your Echo Show's Home screen now includes an icon of a pointing finger tapping the screen. Tap that icon and you see the Tap to Alexa screen, shown in Figure 10-1. Note that you can swipe left to see a second page of tiles.

FIGURE 10-1:
The default first page of the Tap to Alexa screen.

What happens when you select a tile depends on the underlying task. Tapping the Weather tile is the same as saying, "Alexa, tell me the weather," so you immediately hear Alexa giving you the current weather conditions and forecast. Similarly, tapping the Music tile is the same as saying, "Alexa, play some music," so you hear a random playlist. Other tiles such as Traffic, News, and Joke also invoke Alexa immediately.

Some tiles represent tasks where Alexa requires more information from you. For example, tapping the Timer tile displays a screen that presents you with several options for the length of the timer. Similarly, tapping the Shopping List tile displays a screen that enables you to enter the item you want Alexa to add to the list.

The Tap to Alexa screen is also home to the following features:

>> **Add New tile:** Tap this tile to create your own Tap to Alexa tiles. When you tap Add New, the on-screen keyboard appears, which you use to enter the command you want Alexa to carry out. You're then prompted to select an icon and enter a label for the tile. When you're done, your new tile appears in the Tap to Alexa screen (starting on the second page) using the icon and label you specified.

>> **More options icon:** Tap this icon — the three vertical dots in the upper-right corner of the screen — to customize the Tap to Alexa screen. You get three options:

- **Edit:** Rearrange the Tap to Alexa tiles by dragging and dropping them. You can also delete a tile you don't use by tapping the X that appears with the tile.

- **Layout:** Set the density of the Tap to Alexa tiles: Low (three tiles), Medium (six tiles arranged in two rows), Default (eight tiles arranged in two rows), or High (ten tiles arranged in two rows).

- **Reset all to default:** Reverses all your changes and reverts the Tap to Alexa screen to its original configuration.

Texting without speech

Alexa's feature that enables you to speak a text message and have it sent to someone in your Contacts list is convenient and easy. Well, it's convenient and easy *if* you can say the "Alexa, send a text" command and speak your message. If you can't, are you out of luck? Nope, not if you have an Alexa device with a screen. The Echo Show supports a feature that enables you to send texts (as well as place phone calls) by tapping the screen.

First, follow these steps to enable the Communication without Speech feature:

1. **Swipe down from the top of the screen.**

2. **Tap Settings.**

3. **Tap Accessibility.**

4. **Tap the Communication without Speech switch to on.**

With the Communication without Speech switch enabled, when you swipe down from the top of the screen you see a new communicate icon — a speech bubble — which represents the Calling & Message feature.

Here's how to send a text by tapping the screen:

1. **Swipe left from the right edge of the screen.**

2. **Tap the speech bubble (Communicate icon).**

   The Communicate screen appears.

3. **Tap Show Contacts.**

   Alexa displays a list of the contacts eligible for messaging.

4. **Tap the person to whom you want to send your message.**

5. **Tap the Message icon.**

   Alexa displays an on-screen keyboard.

6. **Type your message.**

7. **Tap Done.**

   Alexa sends your message.

Checking Out Alexa's Features for the Hearing Impaired

"Alexa, speak slower."

If your hearing has deteriorated over the years, or if you have a hearing impairment in one or both ears, hearing Alexa's responses and enjoying music and videos can be a challenge. Fortunately, help is at hand. Alexa has a few tools that you can configure to help with or work around your hearing issues. The next few sections give you the lowdown on the hi-fi.

Making Alexa more hearing accessible

Alexa and the Alexa app come with a few features that you can customize or turn on to work around hearing impairments:

>> The light ring provides a visual indicator of the status of your Echo device (see Chapter 3).

>> For Alexa devices with a screen, on-screen messages and icons provide visual indications of the device's status.

>> Alexa devices have either physical buttons you can press to adjust the volume or a volume ring (see Chapter 3).

>> You can use the Alexa app to adjust the volume for alarms, timers, and notifications (see Chapter 7).

>> You can configure Alexa to display closed-captioning on videos and Alexa captioning on Alexa's responses (see "Enabling closed-captioning" and "Enabling Alexa captioning," later in this chapter).

>> You can connect Alexa with external speakers or headphones, either directly using a cable or wirelessly via Bluetooth (see Chapter 2).

ALEXA WITH HEADPHONES

If you suffer from only minor hearing loss, you may be able to compensate by turning up the volume of your Alexa device. However, this may be impractical if other people are nearby and you don't want to disturb them. A better solution is to use headphones. Because the sound from the headphones takes a shorter and more direct path to your ear, it can make the sounds sharper and easier to discern — and it has the added advantage of not disturbing anyone within earshot.

If you decide to invest in some headphones, here are a few pointers to bear in mind:

- If you use an in-the-ear (ITE) hearing aid, look for ear-pad headphones (also called on-ear headphones), which rest on your ears.

- If you use a behind-the-ear (BTE) hearing aid, you'll need to move up to full-size headphones (also called full-cup, ear-cup, or over-the-ear headphones), which are large enough to cover not only your ear but also your hearing-aid microphone.

- With any type of hearing aid, you need to guard against *feedback*, where amplified sounds from your hearing aid leak out and bounce off the headphones back into the hearing aid. The cycle repeats until a painful feedback squeal is emitted. To prevent feedback, get headphones that use foam ear pads, which reduce the chance of sound being reflected into the hearing aid.

- If your hearing aid comes with a *telecoil mode* (which enables the hearing aid to process sounds sent electromagnetically), be sure to get telecoil-compatible headphones (which broadcast sounds electromagnetically).

- Consider getting noise-canceling headphones, which virtually eliminate background noises to let you hear just the sounds from the headphones.

TIP

If your hearing challenges mean that Alexa speaks too quickly for you to understand, try slowing Alexa's speaking rate by saying "Alexa, speak slower." See Chapter 3 for the details on adjusting Alexa's speaking rate.

Enabling closed-captioning

If you have an Alexa device with a screen, the closed-captioning feature overlays text transcriptions of the voice track in a TV show, movie, or video. Follow these steps to enable and customize closed-captioning:

1. **Swipe down from the top of the screen.**

2. **Tap Settings, and then Accessibility.**

3. **Tap the Closed Captioning switch to on.**

    In some locations, the switch is named Subtitles instead of Closed Captioning.

4. **Tap Closed Captioning Preferences.**

    In some locations, the command is named Subtitle Preferences instead of Closed Captioning Preferences.

    Alexa displays a long list of settings that enable you to customize the appearance of subtitles. *This Is a Preview* appears at the top of the screen to show you what your custom subtitles will look like.

    Example settings include Text Size, Text Color, Font, and Text Background Color.

5. **For each setting you want to customize, tap the setting, use the screen that appears to adjust the setting to your liking, and then tap the back icon (<) to return to the settings.**

TIP

If you end up with closed-captioning subtitles that look downright awful, you can start over again quickly by scrolling to the bottom of the settings and then tapping the Reset to Defaults command.

Enabling Alexa captioning

On an Alexa device with a screen, the Alexa captioning feature overlays text transcriptions — called *subtitles* — of Alexa's responses to your voice commands. Follow these steps to enable and customize Alexa captioning:

1. **Swipe down from the top of the screen.**

2. **Tap Settings and then Accessibility.**

3. **Tap the Alexa Captioning switch to on.**

   In some locations, the switch is named Alexa Subtitles instead of Alexa Captioning.

4. **Tap Alexa Captioning Preferences.**

   In some locations, the switch is named Alexa Subtitle Preferences instead of Alexa Captioning Preferences.

   Alexa displays a screenful of settings for customizing the look of the subtitles.

5. **For each setting you want to customize, tap the setting, use the screen that shows up to adjust the setting to your liking, and then tap the back icon (<) to return to the settings.**

TIP

If you make a mess of the Alexa captioning subtitles, reboot by scrolling to the bottom of the settings and then tapping Reset to Defaults.

Enabling Alexa's Features for the Vision Impaired

Those of you who are no longer spring chickens (or even summer chickens, for that matter) know one thing for certain: The older you get, the worse your eyesight becomes. Sure, you can ramp up your eyeglass prescription or invest in extra-strength reading glasses, but even that may not be enough when it comes to reading text and deciphering icons on your Alexa device screen. And, of course, if your eyesight problems go beyond relatively simple afflictions such as farsightedness or astigmatism, a change of eyewear isn't going to help you make sense of what's happening on the device's screen.

Whatever the source of your visual challenges, you can't work with your Alexa device (or the Alexa app) if you can't see what Alexa is trying to show you on-screen. Fortunately, a number of tools are available for enlarging screen items, changing other features to improve visibility, and even hearing audio transcriptions of what's on the screen.

Making Alexa more vision accessible

Both your Alexa device and the Alexa app come with a few features that you can tweak or activate to work around vision problems:

» Alexa devices have either physical buttons you can press to adjust the volume or a volume ring (see Chapter 3).

» You can use the Alexa app to adjust the volume for alarms, timers, and notifications (see Chapter 7).

» The Alexa app supports the vision accessibility features — such as dynamic type sizes and high contrast — available on your smartphone:

  • *Android:* Choose Settings ⇨ Accessibility ⇨ Visibility Enhancements.

  • *iOS (or iPadOS):* Choose Settings ⇨ Accessibility and adjust the settings in the Vision section.

» If your Alexa device's white text on a black screen is hard to read, you can invert the colors. Swipe down from the top of the screen, choose Settings ⇨ Accessibility, and then tap the Color Inversion switch to on.

» You can operate the Alexa app using a screen reader:

  • *Android:* Choose Settings ⇨ Accessibility ⇨ Screen Reader and then tap the Voice Assistant switch to on.

  • *iOS (or iPadOS):* Choose Settings ⇨ Accessibility ⇨ VoiceOver and tap the VoiceOver switch to on.

» You can navigate your Alexa device screen with a screen reader (see "Navigating with VoiceView," next).

» You can zoom in on your Alexa device screen (see "Zooming in with Screen Magnifier," later in this chapter).

» You can configure your Alexa device to play a sound when Alexa starts processing a request and when it finishes processing that request. In the Alexa app, choose Devices ⇨ Echo & Alexa, tap your Alexa device, and then tap Sounds. Tap the Start of Request and End of Request switches to on, as shown in Figure 10-2.

Navigating with VoiceView

If you have an Echo Show, you can use the VoiceView feature as a screen reader. As you open each screen, VoiceView tells you the name of the screen — you can tap any item on the screen to have the item's name or text read out loud.

To activate VoiceView, follow these steps:

1. Swipe down from the top of the screen and then tap Settings.

2. Choose Accessibility ⇨ VoiceView Screen Reader.

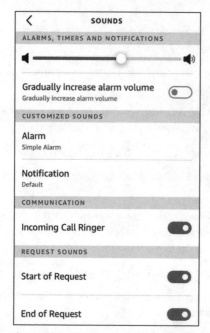

FIGURE 10-2:
Tap Start of
Request and End
of Request to on
to hear sounds
that indicate
when Alexa is
processing your
voice commands.

3. **Tap the VoiceView switch to on.**

   The Alexa device switches into VoiceView mode and a voice says, "VoiceView ready." You may at this point see a tutorial for VoiceView. If so, go ahead and run through the tutorial to learn the basics. When you're done with the tutorial, you can continue with these steps.

   You now see a long list of settings for customizing VoiceView to suit your style.

   While VoiceView is active, how you interact with the screen changes. I talk more about these changes in a bit, but to continue, you need to know about three VoiceView gestures:

   - To hear the name or text associated with a screen item, tap that item.

   - To open a screen item (for example, to run a command), tap the item to select it and then double-tap anywhere on the screen.

   - To scroll a list, use three fingers to swipe down or up.

4. **Customize any of the following preferences as your heart desires:**

   - *Reading Speed:* Sets the speed at which VoiceView reads the screen text.

   - *Verbosity:* Lets you specify which screen elements VoiceView mentions. For example, when the Speak Role switch is on, VoiceView tells you the item type of screen objects, such as buttons and check boxes.

- *Speech Volume:* Sets the VoiceView speech volume relative to the Alexa device's current volume setting. By default, VoiceView uses the device's current volume (the Match Device Volume option), but you can opt for a quieter playback (such as 75 percent or 50 percent of the device volume).

- *Sounds Volume:* Sets the VoiceView sound effects volume relative to the Alexa device's current volume setting.

- *Key Echo:* Specifies how VoiceView confirms what you've input when you're entering text using the on-screen keyboard. By default, VoiceView says each character as you type it and each word as you complete it (that is, it *echoes* the keys you press and the words you complete). You can also opt to have VoiceView echo just characters or just words.

- *Punctuation Level:* Specifies whether and how much punctuation VoiceView includes in its descriptions.

- *Identify Capital Letters:* Specifies how VoiceView indicates uppercase letters.

- *VoiceView Tutorial:* Runs a tutorial that gives you a quick lesson in how to use VoiceView.

With VoiceView activated and configured, you can use the gestures outlined in Table 10-1 to navigate your Alexa device.

TABLE 10-1   **Navigating with VoiceView Gestures**

Task	VoiceView Gesture
Open the status bar.	Swipe down from the top of the screen with three fingers.
Display Alexa actions.	Swipe left from the right edge of the screen with three fingers.
Go to the next page or previous page.	Swipe left or right with three fingers.
Explore the screen.	Drag a finger over items on the screen to have them read aloud.
Go to the next item.	Swipe right using one finger.
Go to the previous item.	Swipe left using one finger.
Open an item.	Tap the item and then double-tap anywhere on the screen.
Increase granularity.	Swipe up and then down using a single motion. This allows you to step through items by either word or character.
Decrease granularity.	Swipe down and then up using a single motion.

(continued)

TABLE 10-1 *(continued)*

Task	VoiceView Gesture
Go to the next item by granularity.	Swipe down.
Go to the previous item by granularity.	Swipe up.
Type on the keyboard.	Move one finger slowly over the keys to hear them read to you. When you hear the letter you want to type, lift your finger from the keyboard to enter the letter.
Scroll vertically.	Swipe up or down with three fingers.
Scroll horizontally.	Swipe left or right with three fingers.
Stop speech.	Single-tap with two fingers.
Read all from selected item.	Swipe down using two fingers.
Read all from first item.	Swipe up using two fingers.
Start or stop media.	Double-tap using two fingers.
Mute or unmute speech.	Double-tap using three fingers.
Adjust a slider.	Tap the slider to select it, and then swipe right and then left in a single motion to increase, or left and then right in a single motion to decrease. Alternatively, double-tap and hold, and then slide your finger to adjust the control.
Enter or exit Learn mode for gesture practice.	Double-tap using four fingers.
Go to the first item on the screen.	Tap the upper half of the screen using four fingers.
Go to the last item on the screen.	Tap the lower half of the screen using four fingers.

Zooming in with Screen Magnifier

You may find that although you can make out most of the items on your Alexa device screen, the occasional icon or bit of text is just too small to decipher. You can always grab a nearby magnifying glass to get a closer look at the section you can't make out, but Alexa offers an electronic version of the same thing. It's called, appropriately enough, Screen Magnifier, and it enables you to zoom in on a portion of the screen.

Screen Magnifier is on by default, but you may want to run through these steps just to make sure:

1. **Swipe down from the top of the screen and then tap Settings.**
2. **Tap Accessibility.**
3. **Tap the Screen Magnifier switch to on.**

To give Screen Magnifier a whirl, use the gestures listed in Table 10-2.

TABLE 10-2   **Screen Magnifier Gestures**

Task	Screen Magnifier Gesture
Zoom in on the screen.	Triple-tap the screen.
Pan a zoomed screen.	Drag two fingers around the screen. Be sure to keep your two fingers slightly apart for panning to work.
Change the magnification level of a zoomed screen.	To zoom out, place two fingers on the screen and pinch them together; to zoom in, place two fingers on the screen and spread them apart.
Zoom out to normal magnification.	Triple-tap the magnified screen.
Temporarily zoom in on the screen.	Triple-tap the screen but leave your finger on the screen after the third tap. You can then pan the screen by dragging your finger.

Using Show and Tell to identify items

If you're blind or have very low vision, you might not be able to identify items that have the same size or shape. For example, you might have a canned good of some kind, but without being able to read the label, you can't tell what the can contains. Fortunately, Alexa has a handy Show and Tell feature that can help you identify a few household objects, including grocery items.

To give Show and Tell a try, hold the item in front of your Echo Show camera, about a foot away from the screen. Then say the following:

"Alexa, what am I holding?"

Alexa trains the camera on the product. If Alexa doesn't recognize what you're holding right away, it prompts you to move the product from side to side and to show different angles or sides of the product.

4
Controlling Your Smart Home

IN THIS PART . . .

Learn how you can extend Alexa's hands-free convenience to your smart-home devices.

Use Alexa to control every type of smart-home gadget, including lights, switches, thermostats, cameras, and TVs.

Make your smart home a veritable genius home by setting up device groups, multifunction routines, and applets that enable Alexa to work with any home automation doodad.

IN THIS CHAPTER

» Finding out how smart smart-home devices really are

» Getting your smart-home devices connected

» Telling Alexa about your smart-home knickknacks

» Managing your increasing horde of smart-home gadgets

» Using Alexa voice commands to control smart-home stuff

Chapter **11**

Setting Up Your Smart Home

Technology's tall-forehead types seem to be obsessed with the word *smart*. It began in earnest in the early 2000s when simple flip phones were replaced by powerful new devices that could surf the web, send email, and much more. Cellphones gave way to smartphones. A few years later, simple time-only watches were replaced by powerful new devices that could count your steps, provide directions, and much more. Wristwatches gave way to smartwatches.

What makes these devices smart depends on who you ask. Some folks say it's the impressive list of tasks you can now perform on the devices. Others claim that a device is smart if it can connect to the Internet and be controlled remotely.

Whatever the reason, the *smart* modifier is poised to take over your home. These days, you can't swing a cat without coming dangerously close to a smart lightbulb, smart thermostat, or smart doorbell. Perhaps the ultimate such device is the smart speaker — particularly any of Amazon's Echo devices — because it enables you to control other smart things with voice commands.

It's this idea of using Alexa to control nearby smart-home devices that forms the heart of this chapter. In the pages that follow, you explore this brave new world of smart-home stuff, learn how to get it all set up, and then dig into the specifics of using Alexa to control your new smart castle. This is going to be a fun one.

What's So Smart about a Smart Home?

Unless you're under 10 years old, you probably grew up in a home that was, well, dumb. You turned on lamps with a switch (or perhaps a clap), the thermostat told you just the current temperature, and the only task you could automate was setting your alarm clock. All that seemed perfectly normal at the time, but that dumb home is starting to look quaint when placed next to the modern idea of a smart home.

What, exactly, do people mean when they add the *smart* adjective to the word *home?* The simple — and not all that helpful — definition of a *smart home* is a home that contains one or more devices that enhance your home life in some way. That word *enhance* is vague, I know, but it's really the key to everything. How does a smart-home device enhance your home life? It comes down to three things:

>> **Convenience:** You operate most dumb-home devices manually, meaning you have to walk up to the device and then throw a switch or adjust a dial. If someone's at the door, the only way to see who's there is to open the door or peek through the peephole. By contrast, you operate smart-home devices from a distance by using an app or a voice assistant such as Alexa. If someone's at the door, your smart security camera lets you see who's there by using an app or an Alexa device with a screen.

>> **Information:** Dumb-home devices tell you either nothing about themselves or just the bare minimum. A dumb dimmer tells you nothing about the current light level, while a dumb thermostat shows only the current temperature. By contrast, smart-home devices are bristling with information — such as current settings, status indicators, and power usage — that gets relayed to an app or device for easy reference.

>> **Automation:** Dumb-home devices just sit there until you do something. A dumb lamp goes on when you flip the switch and will stay on until you flip the switch back. A dumb thermostat will keep the house at the temperature you set, no matter what the time of day. By contrast, smart-home devices can be programmed. You can program a smart lamp to turn on and off automatically at specified times. You can program a smart thermostat to use your preferred temperature during the day, and to use an energy-saving lower or higher temperature overnight.

Yes, some smart-home stuff is a solution in search of a problem. A smart water bottle that tells you when it's time to take a drink and a smart hairbrush that lets you know when you're not brushing correctly are among the dumber smart devices. On the other hand, even something as basic (in the smart-home world, anyway) as programming when your lights go on and off can save you money by reducing energy costs and extending bulb life, and make your home more secure by making it look occupied even when you're not there.

Working with smart-home devices

One of the most confusing things about getting into the smart-home scene is understanding how you work with those smart devices that you bring into your home. How do you control them? How do you get information from them? How do you automate them?

That confusion exists because there isn't just one way to work with any smart-home device. There are, in fact, three ways:

>> **Using an app:** The simplest way to work with a smart-home device is to use the manufacturer's app on your smartphone or tablet. With the app installed, after you plug in and turn on the device, the app will recognize the device (I talk about how this happens in the next section) and set it up. You can then use the app to control the device (say, by turning it on and off), get information about the device, and automate the device in some way. If you add a second device from the same manufacturer, you can use the same app to control that new device, as well. (Although some manufacturers offer separate apps for each of their devices.) However, if you add a device from another manufacturer, you need that company's app to control the new device.

>> **Using a device controller, such as Alexa:** The problem with app-based device control is that as you add smart devices to your home, the number of apps you have to juggle gets larger and larger. Two or three apps is no big deal, but a dozen smart-home apps is unwieldy. The solution is to bypass the apps in favor of a *smart-home controller,* which is a device that can recognize and work with multiple devices from multiple manufacturers. An Alexa device such as an Echo or Echo Show is an example of such a controller. For this to work, you must make sure you get smart-home devices that are compatible with your controller.

>> **Using a smart-home hub:** A smart-home controller lets you bypass an app to control a device, but you still need the app to enable the controller to find the device and to give the controller permission to work with the device. That means the setup for many devices is complex and time-consuming. The solution is to eschew the controller in favor of a *smart-home hub,* which is a device that uses

special networking technology (again, see the next section) to automatically recognize and connect with compatible smart-home devices. Many smart-home device manufacturers also make hubs; the fourth-generation Echo, third-generation Echo Show 10, and Echo Studio are examples of devices that have built-in smart-home hubs.

WARNING

If you want to use Alexa voice commands to control your smart-home devices, you need to be careful because not every device is Alexa-friendly. If you're shopping on Amazon, look for "Works with Alexa" in the item name or description. On the Amazon product page, look for the Amazon Certified badge that guarantees the device can be controlled by Alexa.

Understanding smart-home connections

Another source of confusion (besides how you work with devices, as I talk about in the preceding section) is how smart-home devices communicate. You already know you can send commands to a device and get information from a device, but what mechanism is used to send and receive this data? Because nothing is ever simple with home automation, there are not one, not two, but *three* mechanisms by which smart-home devices can communicate:

» **Wi-Fi:** Most smart-home devices have built-in Wi-Fi hardware, meaning that you can get these devices connected to your home Wi-Fi network, which enables them to work with an app, a controller (such as Alexa), or a hub connected to the same network. Getting the smart-home device on your Wi-Fi network usually involves three steps:

1. The smart-home device creates a temporary Wi-Fi network.

2. You connect your mobile device to the smart-home device's network. This gives the app, controller, or hub access to the device to configure it.

3. You use the app, controller, or hub to connect the smart-home device to your Wi-Fi network.

With newer Amazon Echo devices, the setup procedure now skips Step 2.

» **Bluetooth:** Many smart-home devices — particularly headphones and speakers — come with Bluetooth networking, which enables a controller or hub to connect to (or *pair* with) the device directly, thus avoiding the convoluted Wi-Fi setup steps.

» **Wireless mesh network:** In this type of wireless network, all the devices — they're called *nodes* — can communicate with each other directly. Crucially, mesh network nodes can connect to each other without requiring credentials (as in a Wi-Fi connection) and can make themselves available to multiple devices (unlike a Bluetooth connection, which can exist only between two devices). Not having to jump through these hoops means that a smart-home hub that's compatible with the mesh network can automatically locate and connect to any network nodes it finds. The *compatible* part of that last sentence is important because many different wireless mesh technologies exist. The two most popular are Zigbee and Z-Wave, so whatever smart-home hub you choose, your devices should also support the same type of mesh network. The fourth-generation Echo, third-generation Echo Show, and Echo Studio support Zigbee networks.

So, just to tie together this section and the preceding one, if you're using one or more manufacturer apps to connect and control devices, you'll almost always use Wi-Fi; if you're using a device controller (such as Alexa), you'll connect and control devices by using Wi-Fi or Bluetooth; if you're using a smart-home hub, you'll connect and control devices by using Wi-Fi, Bluetooth, or a mesh network such as Zigbee.

How does Alexa fit in?

You probably already know the answer to this section's question: Alexa is an example of a smart-home device controller, which means you can use an Alexa device to connect to and operate smart-home devices using either your home's Wi-Fi network or a direct Bluetooth connection. In most cases, the connection procedure involves two steps:

1. **For a Wi-Fi connection, use the manufacturer's app to give Alexa permission to connect to the device.**

2. **Enable the manufacturer's device skill to update Alexa with the voice commands needed to operate the device.**

I go into both steps in more detail later in this chapter.

Do you need a smart-home hub?

Because you're reading this book, you're probably using Alexa to control your smart-home devices, and you may be wondering whether you need to upgrade your smart home with a hub. The beauty of a hub is that it locates and connects to compatible mesh-network devices without your having to lift a finger (other than to tap the hub's Locate Devices command).

A smart-home hub is also the way to go if you have some smart devices that aren't compatible with your Alexa device. For example, if you have an older Echo, Echo

Dot, or Echo Spot, Alexa can't work with mesh-network devices such as Zigbee and Z-Wave. However, if you install a smart-home hub that supports these networks, Alexa will be able to control at least some devices on those networks.

If you think you need a hub, there are two ways to go:

>> **Buy an Alexa device that has a built-in hub, meaning a fourth- generation Echo, a third-generation Echo Show, or an Echo Studio.** In this case, you must make sure that your smart-home devices support Zigbee mesh networking.

>> **Buy a third-party hub and then connect it to a non-hub Alexa device, such as an older Echo, Echo Dot, or Echo Spot.** This approach gives you a bit more flexibility because, after you install the hub's Alexa skill, Alexa can now control any device compatible with the hub (which, for most hubs, means Wi-Fi, Bluetooth, Zigbee, and Z-Wave devices, among others).

REMEMBER

If you don't want to spend money on a hardware hub, you can get similar results by going the software route. A service named IFTTT (If This Then That) lets you create small scripts called *applets* that enable you to control smart-home devices whatever their connection. I talk more about IFTTT in Chapter 12.

Setting Up Your Smart-Home Devices

Assuming you have some smart-home devices newly delivered, it's time to figure out how to get those devices set up so that you can control them through Alexa using voice commands. The next few sections take you through the setup procedures for Wi-Fi and Zigbee devices. (I cover connecting Bluetooth stuff back in Chapter 2.)

Installing a Wi-Fi smart-home device

If your smart-home device is Wi-Fi-friendly, go to your mobile device app store and install the manufacturer's app. Then follow these steps to get your Wi-Fi smart-home device set up in the app:

1. Plug in and, if required, turn on the smart-home device.

2. Open the smart-home device manufacturer's app.

3. **Initiate the procedure for setting up a new device.**

   Look for a command named Add or Add *manufacturer* Device (where *manufacturer* is the name of the company) or just a big plus sign (+).

   The setup routine will tell the device to broadcast its Wi-Fi network.

4. **Open your mobile device's Wi-Fi settings and look for the device's Wi-Fi network.**

   Figure 11-1 shows a collection of Wi-Fi networks that includes WeMo. Insight.03C, which is a network broadcast by a WeMo Insight smart switch.

FIGURE 11-1: Open your Wi-Fi settings and look for the smart-home device's Wi-Fi network. Device network

5. **Tap the device network to connect to it.**

6. **When the connection is complete, return to the device app.**

   The app automatically detects the new network and uses the connection to set up the device. This setup usually involves giving the device a name, as shown in Figure 11-2. You'll often have to set up an account with the manufacturer, as well.

7. **The app will usually ask for your Wi-Fi credentials, which enables the device to connect to and operate over your network.**

   Having the device on your network is also how Alexa discovers and operates the device, so this step is important.

FIGURE 11-2:
You usually have
to run through a
few configuration
chores, such as
giving the device
a silly name.

8. **If you see a notice asking whether you want to upgrade the smart-home device firmware, by all means tap Yes or Allow or Update or whatever option answers in the affirmative.**

   The firmware is internal software that runs the device. Keeping all your smart-home devices updated with the latest firmware is highly recommended because new versions of the software are often needed to patch security holes and improve performance.

Understanding Wi-Fi Simple Setup

The steps I outline in the preceding section mostly deal with getting a Wi-Fi-enabled smart-home device on your home network. The step where you need to connect your smartphone or tablet to the device's temporary network always bothers me because it seems like an imposition. The Amazon engineers must have felt the same way, because they came up with a way to avoid that annoying extra step. It's called Wi-Fi Simple Setup, and it requires two things:

>> An Echo device compatible with Wi-Fi Simple Setup — that is, a second-generation or later Echo, an Echo Plus, an Echo Dot, or an Echo Show — that's already connected to your Wi-Fi network

>> The password to your Wi-Fi network saved to Amazon

If you've checked off both items, setting up a new device that's compatible with Wi-Fi Simple Setup — such as the Amazon smart plug or the AmazonBasics microwave — is either easy or ridiculously easy.

The ridiculously easy setup comes your way if you purchased your Wi-Fi Simple Setup device from Amazon. In that case, Amazon automatically associates the device with your Amazon account, which means that when you plug in the device, it will connect to your network automatically using your saved Wi-Fi password. Now *that's* ridiculously easy!

If you purchased the Wi-Fi Simple Setup device from a retailer other than Amazon, the device won't be associated with your Amazon account, so it can't connect to your network automatically. That's okay, though, because you can still use the Alexa app to add the device: Choose Devices, tap + (add (+), and then tap Add Device.

Discovering smart-home devices using an Alexa skill

If you're not using an Alexa device that includes a smart-home hub, you usually need to upgrade Alexa to work with your smart-home device. You upgrade Alexa by enabling the device manufacturer's Alexa skill. This not only lets Alexa discover the device but also upgrades Alexa with the voice commands that let you operate the device.

REMEMBER

Alexa can locate and connect to some smart-home devices without requiring a skill. For example, Alexa can work with a Philips Hue Bridge to control lights without needing a skill. Say, "Alexa, discover devices," and then press the button on top of the Hue Bridge to put it into pairing mode.

Follow these steps to enable the manufacturer's Alexa skill and discover the manufacturer's smart-home device:

1. **Install the manufacturer's app and use it to get your smart-home device on your Wi-Fi network.**

   See "Installing a Wi-Fi smart-home device," earlier in this chapter, for the details.

2. **In the Alexa app, tap Devices.**

3. **Tap the add icon (+) that appears in the top-right corner.**

4. **Tap Add Device.**

   The Alexa app displays icons for some popular brands and some device categories.

5. **Tap the category that fits your device, and then tap the manufacturer.**

   The Alexa app prompts you to perform the duties I outline in Step 1. You've done all that, so proceed.

6. **Tap Continue.**

The Alexa app opens the information page for the manufacturer's Alexa skill.

7. **Tap Enable and follow the on-screen instructions.**

At this point, what happens next depends on the skill, but you'll usually have to perform one or both of the following:

- *Use the smart-home device app to give Alexa permission to access the device.*

- *Link Alexa to the user account associated with the smart-home device.*

8. **When you're done, tap close (X) to return to the skill page.**

9. **Tap Discover Devices (see Figure 11-3, left).**

The Alexa app uses the manufacturer's Alexa skill to search for available devices. If one or more devices are found, you see a screen similar to the one shown in Figure 11-3, right.

FIGURE 11-3:
Tap Discover
Devices (left) to
see which devices
(if any) are
available (right).

10. **Tap Set Up Device.**

Alexa now takes you through a device setup screen or two, the specifics of which depend on the device type.

11. **Tap Done.**

With a manufacturer's Alexa skill enabled, you can discover new devices by following Steps 1 through 5 and then tapping Discover Devices, or you can ask Alexa to run the following voice command:

"Alexa, discover my devices."

Discovering Zigbee smart-home devices

If you have one or more smart-home devices that support Zigbee and you have a fourth-generation Echo, a third-generation Echo Show, or an Echo Studio, the built-in hub in your Echo will automatically discover your devices.

Before continuing, make sure your Zigbee devices are in pairing mode by turning them off and then back on again.

The easiest way to get Alexa to find your Zigbee devices is to issue the following voice command:

"Alexa, discover my devices."

Otherwise, you can use the Alexa app as follows:

1. **Tap Devices.**

2. **Tap the add icon (+) that appears in the top-right corner.**

3. **Tap Add Device.**

   The Alexa app displays icons for some popular brands and some device categories.

4. **Tap the category that fits your device and then tap the manufacturer.**

   If you don't see the manufacturer, tap Other.

5. **Tap Discover Devices.**

   The Alexa app scans for nearby pairable Zigbee devices. If it finds one or more devices, you see a screen similar to the one that was shown in Figure 11-3.

6. **Tap Set Up Device and then follow the setup instructions that appear.**

7. **Tap Done.**

   Your Zigbee devices are ready to go.

Managing Smart-Home Devices

After you have one or more smart devices installed and connected to Alexa, you can start controlling those devices with voice commands. Sweet! I get into that a bit later, but for now you should know a few techniques for managing your smart-home devices.

Viewing your smart-home devices

The easiest way to keep track of what smart-home devices you have installed is to use the Alexa app. Tap the Devices icon to open the Devices screen, as shown in Figure 11-5. The top part of the Devices screen is divided into categories as follows:

>> **Echo & Alexa:** Displays a list of your Echo devices and Alexa gadgets.

>> **Type:** You see a separate icon for each type of device, such as the lights, plugs, and switches types shown in Figure 11-4.

>> **All Devices:** Displays every device connected to Alexa.

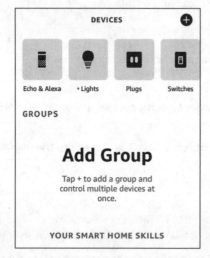

FIGURE 11-4:
Use the Devices screen to keep track of your installed smart-home devices.

REMEMBER

If you see a little green dot next to a device category, Alexa has recently discovered or connected to a new device in that category.

To view information about (and, as I show in the next few sections, make changes to) a device, tap the device category, tap the device, and then tap the device settings icon (gear) in the upper-right corner. Figure 11-5 shows an example of the Settings screen that appears.

FIGURE 11-5:
Use the Settings
screen to get
device info and
perform a few
device-related
tasks.

Most of the data you see on the Settings screen is straightforward (or explained in the next few sections). However, I'd like to point out the Connected Via setting, which tells you how Alexa is connected to the device. Depending on the device, you'll see one of the following three connection types:

>> **Alexa skill:** If Alexa used a manufacturer's skill to connect to the device, the Connected Via setting shows the manufacturer's name.

>> **Network:** If Alexa connected to the device using Wi-Fi, Bluetooth, or Zigbee, the Connected Via setting shows the network type.

>> **Hub:** If Alexa connected to the device through a smart-home hub, the Connected Via setting shows the name of the hub.

Renaming a smart-home device

You usually name a smart-home device using the manufacturer's app, but you can also follow these steps to rename the device using the Alexa app:

1. **Tap Devices.**

2. **Tap the device type (or All Devices) and then tap the device you want to rename.**

   If your All Devices list is long, an easier way to locate your device is to tap the appropriate device type at the top of the Devices screen.

3. **Tap the device settings icon (gear) in the top-right corner of the screen.**

4. **Tap Edit Name.**

   The Alexa app opens the device name for editing.

5. **Edit the name of the device.**

TIP

   You'll be saying some device names over and over, so make sure you give any frequently used device a name that's short, easy to pronounce, and memorable.

6. **Tap Done.**

   The Alexa app renames the device.

Changing the device type

Each device's Settings screen includes a Type setting, which tells you the device type, such as Light, Plug, or Switch. For most devices, the type is set in the digital equivalent of stone, but some devices can change type. For example, a smart plug is, of course, a plug, but if you connect, say, a lamp, isn't the device now a light instead of merely a plug?

Okay, sure, but why does that matter? It matters because, as I show in Chapter 12, you can set up device groups that you can control as a unit. If you have, say, several smart lights in that group, but you also want to include a lamp connected to a smart plug, it's easier if you tell Alexa to treat the smart plug as a light, instead of just a plug.

For devices that can take on different types, follow these steps to change the type using the Alexa app:

1. **Tap Devices.**

2. **Tap the device type (or All Devices) and then tap the device you want to change.**

3. **Tap the device settings icon (gear) in the top-right corner of the screen.**

4. **Tap Type.**

   The Alexa app displays the Select Type screen, an example of which is shown in Figure 11-6.

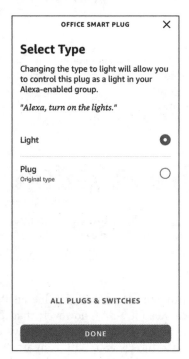

FIGURE 11-6:
On supported
devices, use the
Select Type
screen to change
the device type.

5. **Tap the device type you want to use.**

6. **Tap Done.**

Alexa now treats the device as the type you selected.

Disabling or deleting a smart-home device

If you're not using a smart-home device temporarily, you can disable it to reduce clutter in your device lists. Alternatively, if you're never going to use a smart-home device again, you can delete it permanently from Alexa. Follow these steps in the Alexa app to disable or delete a device:

1. **Tap Devices.**

2. **Tap the device type (or All Devices) and then tap the device you want to disable or delete.**

3. **Tap the device settings icon (gear) in the top-right corner of the screen.**

4. **Tap the Enabled switch to off.**

   Alternatively, if you prefer to remove the device completely, tap Delete (the trash can icon in the top-right corner of the screen), and then tap Delete when the Alexa app asks you to confirm.

   The Alexa app disables or deletes the device.

Working with Smart-Home Devices

With your smart-home devices plugged in, turned on, connected to Alexa, and sensibly named, you're ready to reap the harvest of all that labor: controlling those devices through Alexa using voice commands. Don't let all that power go to your head!

REMEMBER

In the sections that follow, I outline the Alexa voice commands that are generally available for each type of device. Keep in mind, however, that the ways you can control a smart-home device through Alexa are almost always only a subset of what you can do using the manufacturer's app. With a smart plug, for example, Alexa can only turn the device on or off, but the manufacturer's app will usually let you schedule on/off times, turn the plug off automatically after a set time, and more.

Controlling a smart-home device

Before getting to the meat of this section, you should know that you can use three methods to control a smart-home device:

>> **Voice commands:** This is how you'll operate most of your smart-home devices, and the rest of this section takes you through the most common voice commands for a selection of smart-home devices.

>> **Alexa app:** If you have your Alexa device microphone turned off, you can still use the Alexa app to control your smart-home devices. Tap Devices, tap the device type (or All Devices), and then tap the device you want to mess with. The screen that appears contains the controls you can use. For example, Figure 11-7 shows the device screen for a smart lightbulb, which includes two controls: a button for turning the device on and off and a slider for setting the brightness.

FIGURE 11-7:
The device
control screen for
a smart lightbulb.

>> **Alexa device with a screen:** Swipe left from the right edge of the screen, tap Smart Home, and then tap the icon for the device type (such as a bulb icon for your smart lights, plugs, and switches). Note, too, that after you issue a smart-home device-related command to an Alexa device with a screen, you see some device controls on the screen for a few seconds.

Working with device scenes

Another "device" that Alexa may discover when you shout out the "Alexa, discover my devices" command is a scene. In the smart-home world, a *scene* is a custom collection of devices and device settings. For example, if a room has four smart lightbulbs, a scene may include two of those bulbs and set them to 50 percent brightness. Many smart-home devices come with one or more prefab scenes, which you can check out in the manufacturer's app. In most cases, that app also gives you a way to create your own scenes.

When you enable the manufacturer's skill and ask Alexa to discover devices, Alexa checks in with the manufacturer's app and connects not only to any devices defined in the app but also to any scenes defined in the app. If Alexa discovers at least one scene, the Devices screen in the Alexa app sprouts a new Scenes tab, as shown in Figure 11-8, left. Tap Scenes to get a list of the available scenes, as shown in Figure 11-8, right.

SCENES

Bright in Living Room

Bright in Office

Dimmed in Living Room

Dimmed in Office

Nightlight in Living Room

Nightlight in Office

Discover Scenes

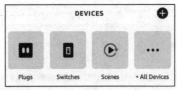

DEVICES

Plugs Switches Scenes + All Devices

FIGURE 11-8:
Tap Scenes (left)
to see what's
available (right).

To activate a scene, use the following voice command:

"Alexa, turn on *scene.*"

Replace *scene* with the name of the scene you want to run, where the name you say must be one of the names in the Scenes screen (such as "Bright in Living Room" shown in Figure 11-8, right).

TIP

A typical scene uses a moniker such as "*name in room,*" where *name* is the label for the scene settings and *room* is the space to which the scene applies (that is, where the devices are located). This means that you can use voice commands such as "Alexa, turn on *room*" or "Alexa, turn off *room*" to control the devices in the location given by *room*.

Whenever you create or change a scene, let Alexa know by using the Alexa app to open the Scenes screen and tap the Discover Scenes button.

TIP

If you delete a scene in the smart-home device app, Alexa doesn't usually remove that scene, so you need to do it manually. Use the steps I outline earlier in the "Disabling or deleting a smart-home device" section.

Turning smart plugs on and off

If you are curious about smart-home technology but don't want to spend a ton of money or time, a smart outlet — most often called a smart plug — is the way to go. A *smart plug* is an electrical outlet that you can control with voice commands. The smart outlet plugs into a regular electrical outlet for power and then you plug a non-smart device — such as a lamp or coffeemaker — into the smart outlet. Voilà! You now have voice control over the dumb device.

TIP

Do you have a bunch of nearby dumb devices that you want control via Alexa? In that case, instead of getting multiple smart plugs, buy a single smart power strip.

Note, however, that "control" here just means turning the device on and off using the following voice commands:

>> "Alexa, turn *device name* on."

>> "Alexa, turn *device name* off."

Replace *device name* with the name you gave to the smart plug using either the manufacturer's app or the Alexa app (see "Renaming a smart-home device," earlier in this chapter).

Working with smart lights

Another easy and relatively inexpensive way to get your smart-home feet wet is with a smart lightbulb or two. You can buy a smart bulb for less than $20, and installing it is as easy as changing any regular lightbulb. You can also get smart bulbs that change brightness without a separate dimmer switch and that can display different colors.

What if you have a large collection of lights in, say, your kitchen or living room? Swapping out all those dumb bulbs for smart versions would cost a fortune, so a better choice is a smart light switch that you can turn on and off via Alexa. For more control, you can get a smart dimmer switch that enables you to control the brightness with voice commands.

WARNING

Although a smart light bulb is easy to install, a smart light switch is another matter because it must be wired to your home's electrical system. Unless you really know what you're doing, hire an electrician to do the installation for you.

Here are the voice commands to use to turn a smart lightbulb or light switch on or off:

>> "Alexa, turn *device name* on."

>> "Alexa, turn *device name* off."

For dimmable smart lights (or smart dimmer switches), use any of the following voice commands:

>> "Alexa, brighten *device name*."

>> "Alexa, dim *device name*."

>> "Alexa, set *device name* brightness to *number* percent."

For smart lights that support different colors, use these voice commands:

>> "Alexa, set *device name* to warm white."

>> "Alexa, set *device name* to cool white."

>> "Alexa, set *device name* to *color*." (For example, "Alexa, set Chill Room to blue.")

Controlling smart thermostats

A smart thermostat gives you voice control over not only the thermostat mode — for example, switching between heating and cooling — but also the temperature setting. You can also interrogate a smart thermostat to learn the current temperature setting.

Smart thermostats that work with Alexa are a little on the pricey side — you can easily spend over $200 — and require installation by someone qualified to work with electrical connections, so be sure the benefits of being able to raise or lower the temperature with your voice are worth it.

After you've installed a smart thermostat, you can control it using the following voice commands:

>> "Alexa, set *device name* to cool."

>> "Alexa, set *device name* to heat."

>> "Alexa, set *device name* to auto."

>> "Alexa, set *device name* to off."

- >> "Alexa, set *device name* to *number* degrees."

- >> "Alexa, increase *device name* by *number* degrees."

- >> "Alexa, decrease *device name* by *number* degrees."

- >> "Alexa, what is *device name*'s temperature?"

- >> "Alexa, what is *device name* set to?"

Keeping an eye on things with smart cameras

It's a fact of life that we can only ever be in one place at a time, but that doesn't stop us from wanting to know what's going on elsewhere. You might want to monitor your sleeping child, see what your dog or cat gets up to while you're not around, or get a look at who just rang the front doorbell.

You can do all these things and more with a smart camera connected to an Alexa device that has a screen. Smart cameras come in both outdoor versions and indoor versions. Outdoor cameras are usually called security cameras or surveillance cameras. You can also get a *smart doorbell*, which has a built-in camera that activates whenever someone rings the bell. Alexa even supports smart doorbells with two-way communications, so you can talk to your visitor without getting up. Indoor smart cameras include specialized versions for monitoring babies, nannies, and pets.

If you have an Alexa device with a screen, you can show the smart camera's feed right on your screen by saying the following voice command:

"Alexa, show *device name*."

REMEMBER

Most smart cameras have a maximum time they'll display the feed, such as 30 minutes. If you bump up against that ceiling, hide the feed and then show it again to get a fresh start.

To turn off the camera feed, issue the following voice command:

"Alexa, hide *device name*."

Other types of smart-home gear

The smart-home gadgets I cover in the last few sections are the most common device types, but other types of home gear are slapping on the *smart* label faster than you can say "Alexa." Here's a quick summary (in no particular order) of what's out there that's Alexa–friendly as I write this:

>> **Smart appliances:** These are kitchen appliances that you can operate with voice commands. Amazon's new Alexa-friendly microwave is the latest addition to this category.

>> **Smart garage door openers:** You can open and close your garage door with your voice.

>> **Smart locks:** These are door locks that you can control with your voice.

>> **Smart speakers:** Echo isn't the only smart speaker game in town. Big-time audio companies such as Bose, Denon, and Sonos have their own smart speakers that have Alexa built in. If you want top-quality sound, forget Echo and go with a smart speaker from an audio specialist.

>> **Smart sprinklers:** You can control and schedule garden and lawn watering with a smart sprinkler system.

>> **Smart TVs:** This is a television or remote that you can control by voice.

>> **Smart vacuums:** Why push a vacuum yourself when you can have a robot vacuum that you control via voice?

>> **Smart window shade:** You can use voice commands to raise, lower, open, and close your window shade or blinds.

Chapter **12**

Uncovering Some Smarter Smart-Home Techniques

When I first started playing around with home automation in the early 2000s, the state of the art was woeful. Few standards existed, companies used proprietary technologies, the connections were spotty, and the control interfaces (Windows only, of course) were mind-numbingly complicated. So, you can imagine my glee these days when I insert a Wi-Fi Simple Setup plug or a Zigbee light bulb and it's connected, configured, and ready to command within a few seconds.

Today's smart-home stuff still has its complications, as I describe in Chapter 11, but the trend is toward faster and easier discovery and connection. So if your smart-home ambitions are satisfied with plugging in a few gadgets and controlling them through your Alexa device, I for one will not judge. However, you can still learn quite a few techniques to improve your smart home. In this chapter, I fill you in on seven of these take-it-to-the-next-level features. If you're looking to make your smart home even smarter, read on.

Creating Device Rooms and Groups

When you have multiple smart-home devices, you run face-first into the biggest obstacle in the home automation game: You have to turn on (and, later, turn off) every device one at a time. This is a two-for-the-price-of-one problem:

>> You have to remember the name assigned to each device.

>> The more devices you have, the longer it takes to get them all powered up or down.

Oh, it's fun for the first few days, but a week into barking out multiple "Alexa, turn on *device name*" voice commands and you'll be ready to throw your smart-home gear into the nearest dumb trash can.

Fortunately, Alexa feels your pain and offers a two-for-the-price-of-one solution:

>> **Smart-home room:** A collection of devices in a common area, such as your bedroom, living room, hallway, or office. Most rooms are physical rooms, but they don't have to be. A room could be, say, nook in the kitchen, a chill space in the living room, or a home theater area in the basement.

>> **Smart-home group:** A collection of related devices, such as all exterior lights or all the smart plugs on one floor.

A smart-home room or group is a collection of devices that Alexa treats as a unit. When you ask Alexa to turn on a room or group, Alexa handles the previously frustrating and time-consuming chore of turning on each and every device that's in that room or group. You just sit back, sip your kombucha, and watch it all unfold.

Creating a smart-home room or group

If smart-home rooms and groups sound like just the ticket, you can create all the rooms or groups you need by using the Alexa app. Here are the steps to plow through:

1. **Tap Devices.**

2. **Tap + (add) in the upper-right corner.**

3. **Tap Add Group.**

4. **Tap Create a Room or Device Group and then tap Next.**

   The Alexa app prompts you for a room or group name. To help you out, the app displays a long list of common names such as Bedroom and Kitchen.

5. Tap one of the preset names or use the Customized Name text box to enter a name; then tap Next.

The Alexa app now prompts you to select the devices that are part of the room or group.

WARNING

After you tap Next, you may see a notification telling you the name is already in use. This usually means the name is being used by a smart-home device app and has been discovered by Alexa. You'll need to use another name for the group or delete the name from Alexa's All Devices list (see Chapter 11).

6. In the Devices section, select the check box beside each device you want to add to the group (see Figure 12-1, left).

7. In the Scenes section, tap each scene you want to add to the group, and then tap Next.

8. If your location has an Alexa device, select its check box in the Available Alexa Devices section (see Figure 12-1, right).

9. Tap Next and then tap Done.

The Alexa app creates your room or group.

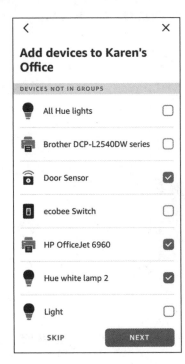

FIGURE 12-1:
Tap each device you want to control via the room or group (left) and then tap the location's Alexa device, if it has one (right).

Controlling a smart-home room or group

With one or more rooms or groups ready for action, you can control any room or group by invoking the following voice commands:

» "Alexa, turn on *room or group*."

» "Alexa, turn off *room or group*."

In both cases, replace *room or group* with the name you gave to your room or group. You can also turn specific device types on and off as follows:

» "Alexa, turn on *type* in *room or group*."

» "Alexa, turn off *type* in *room or group*."

Replace *type* in both commands with the device type you want to control: lights, plugs, and so on.

You can also control a smart-home room or group using the always-handy Alexa app. When you create a room or group, the Alexa app adds it to the Groups section of the Devices screen, as shown in Figure 12-2.

FIGURE 12-2:
The rooms or groups you create appear in the Groups section of the Devices screen.

The Alexa app gives you three ways to control a room or group or its devices:

» **Turn a device type on or off.** On the Devices screen, each room or group includes an icon for every type of device in the group. For example, the

Karen's Office group shown in Figure 12-2 has two icons: one for lights (on the left) and one for plugs (on the right). Tap a device icon to toggle that type of device on and off.

>> **Turn all the room's or group's devices on or off.** On the Devices screen, tap the group to open it; then tap either All On or All Off.

>> **Turn individual devices on or off.** On the Devices screen, tap the group to open it. Then tap a device switch to toggle that device on and off (see Figure 12-3).

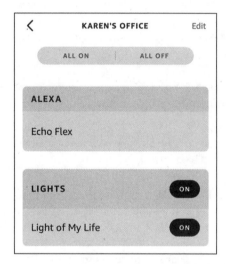

FIGURE 12-3:
Opening a group enables you to control all the devices or individual devices.

Managing a smart-home room or group

If you add new devices to a room or group location, remove devices from that location, or decide you want to modify the room or group name or device set, the Alexa app is the place to go for these room and group management chores. Follow these steps:

1. **In the Alexa app, tap Devices.**

2. **Tap the group you want to edit, and then tap Edit.**

3. **To change the group name, tap the Edit Name link, make your changes, and then tap Next.**

4. **To add a new Echo, device, or scene to the group, select its check box.**

5. **To remove an Echo, device, or scene from the group, deselect its check box.**

6. **Tap Save.**

Alexa in Stereo: Working with Speaker Sets

I have a confession to make: Although I love music and have played in garage bands, I have a tin ear when it comes to sound quality. Sure, I can tell that music sounds awful on a cheap radio and great through mortgage-the-house-expensive speakers. But everything in between those extremes fools my ears. Which is a long-winded way of saying that I like the sound I get out of all my Echo devices, from my big Echo Show 10 with its three speakers all the way down to the little Echo Flexes scattered around the house.

In Chapter 2, I explain that you can get even better sound out of your Echo by connecting wired or wireless external speakers. That can be an expensive way to go, however, so I'm pleased to report that Amazon offers two other roads to better sound:

>> **Stereo:** You can pair two Echo devices to create a stereo effect: One Echo device plays the audio output's left channel, while the other Echo device plays the right channel.

>> **Subwoofer:** You can pair an Echo device with a subwoofer, which is a device that puts out a deep, rich bass sound.

These combinations are called *speaker sets*. Note that these routes aren't either-or. If you have two Echo devices and a subwoofer, you can pair all three devices for the ultimate in Alexa sound systems.

Before getting to the steps, here are some notes to ponder:

>> **For a stereo pair, the two devices you use must be identical: the same Echo model and generation.** For example, you can pair two fourth-generation Echo Dots, but you can't pair a fourth-generation Echo Dot with a third-generation Dot or any generation of Echo or Echo Flex.

>> **Speaker sets work with only music output.**

>> **Your Echo speakers must be online.**

>> **Your Echo speakers must be on the same Wi-Fi network.**

>> **For best stereo sound quality, place your Echo speakers a few feet apart.**

>> **Place your subwoofer on the floor, about a foot from the wall.** If you're also pairing two Echo speakers, place the subwoofer equidistant from each speaker.

Creating a speaker set

Okay, I'm finally ready to reveal the steps to follow to create a speaker set:

1. **In the Alexa app, tap Devices.**

2. **Tap + (add) in the upper-right corner.**

3. **Tap Combine Speakers.**

    The Alexa app displays the Setup screen.

4. **Tap Stereo Pair/Subwoofer.**

    The Alexa app displays a list of your Echo devices.

5. **Tap the devices you want to include in your speaker set.**

    For example, Figure 12-4, left shows two Echo Dots selected for a stereo pairing.

    Note that the Echo Show has the status Not Compatible. That's confusing because it does *not* mean the Echo Show isn't compatible with speaker sets. Instead, it means the Echo Show isn't compatible with the selected Echo Dots for creating a speaker set.

6. **Tap Next.**

    If you're creating a stereo pair, the Alexa app asks you to test the stereo setup, as shown in Figure 12-4, right.

7. **Tap Left and listen for Alexa saying, "Left channel"; tap Right and listen for Alexa saying, "Right channel."**

    If the channels are backward, you can tap Swap Speakers to switch the left and right outputs.

8. **Tap Next.**

    The Alexa app asks you to choose a name for the speaker set.

9. **Tap a suggested name or use the Customized Name text box and type a name.**

10. **Tap Save.**

    The Alexa app sets up the speaker set, which takes a minute or so.

When the pairing is complete, you're sent back to the Devices screen, where you now see a new section called Speaker Groups that includes your speaker set, as shown in Figure 12-5, left.

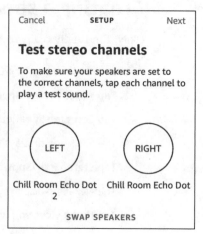

FIGURE 12-4:
Tap the devices you want to include in your speaker set (left), and then, for a stereo pair, select an audio channel for one of the devices (right).

Select speakers to connect
You can select up to two speakers and a subwoofer. All speakers must be online, on the same network and in the same room.

SPEAKERS

Chill Room Echo Dot
Online • LogophiliaB network ✓

Chill Room Echo Dot 2
Online • LogophiliaB network ✓

Echo Show
Not Compatible • LogophiliaB n... ⓘ

Test stereo channels
To make sure your speakers are set to the correct channels, tap each channel to play a test sound.

LEFT   RIGHT

Chill Room Echo Dot   Chill Room Echo Dot
2

SWAP SPEAKERS

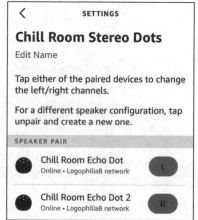

FIGURE 12-5:
Your speaker sets appear in the Devices screen in the Speaker Groups section (left); tap your speaker set to see its settings (right).

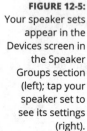

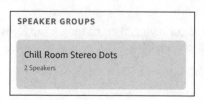

SPEAKER GROUPS

Chill Room Stereo Dots
2 Speakers

‹   SETTINGS

Chill Room Stereo Dots
Edit Name

Tap either of the paired devices to change the left/right channels.

For a different speaker configuration, tap unpair and create a new one.

SPEAKER PAIR

Chill Room Echo Dot
Online • LogophiliaB network   L

Chill Room Echo Dot 2
Online • LogophiliaB network   R

Tap the speaker set to open the Settings screen, shown in Figure 12-5, right. If you want to swap the audio channels, tap any speaker, and then tap the channel you want the speaker to use. The Alexa app automatically assigns the other channel to the second speaker.

Playing music through a speaker set

To use your speaker set as the playback device, include the speaker set in your voice command. Here are some examples:

» "Alexa, play some music on *speaker set name*."

» "Alexa, pause *speaker set name*."

If you want to replay some recent music through your new speaker set, open the Alexa app, tap Play, tap the music you want to hear in the Recently Played section, and then tap your speaker set in the Play On list that appears (see Figure 12-6).

FIGURE 12-6:
Tap some recently played music and then tap your speaker set.

PLAY ON:

Chill Room Stereo Dots

Echo Flex

Echo Show

To control the volume on the speaker set, press the Volume Up or Volume Down control on any device in the set to adjust the volume for both speakers. You can also include the speaker set name in your voice commands, as shown in these examples:

>> "Alexa, turn up the volume on *speaker set name*."

>> "Alexa, mute *speaker set name*."

Streaming Music to Multiple Rooms

If you have multiple Alexa devices and you want to command one device to play music on another, you use a voice command such as this:

"Alexa, play some music on *device*."

Replace *device* with the name of the Alexa device you want to use for playback.

That's pretty sweet, but consider these multi-device musical scenarios:

>> You have multiple Alexa devices in the same room, and you want them all to play the same music.

>> You have multiple Alexa devices in several rooms, and you want them all to play the same music.

>> You have multiple Alexa devices all through your home, and you want to play the same music on every device.

You'd need some kind of high-end audio gear to make any of those scenarios a reality, right? Nope. Alexa can do it all without breaking a digital sweat thanks to one of its most awesome features: *multi-room music.* The idea is that you define a room or group location, such as Second Floor, Downstairs, or Everywhere; add to that group all the Alexa devices you want to play the same music; and then tell Alexa to play music in that location.

TIP

"Hey, wait a minute, Author Boy. What if I want to play, say, two different music streams in two separate groups?" Yep, that scenario's possible, but you can do it only if you have an Amazon Music Unlimited Family plan.

Setting up multi-room music

Here are the steps to follow to make multi-room music happen:

1. **In the Alexa app, tap Devices.**

2. **Tap + (add) in the upper-right corner.**

3. **Tap Combine Speakers.**

4. **Tap Multi-Room Music.**

   The Alexa app prompts you to select the devices you want to connect.

5. **Tap each speaker or speaker set you want to include, as shown Figure 12-7; then tap Next.**

   The Alexa app prompts you for a name. The app displays a long list of common names such as Everywhere, Kitchen, and Downstairs.

6. **Tap one of the existing names or use the Customized Name text box and enter a new name.**

7. **Tap Save.**

   The Alexa app sets up the multi-room speakers, which takes a minute or so.

In the Devices screen, you now see your multi-room music group in the Speaker Groups section, as shown in Figure 12-8.

Playing multi-room music

To use your multi-room music group as the playback device, include the group name in your voice command. Here are some examples:

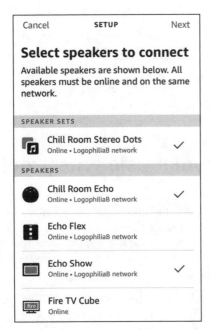

FIGURE 12-7:
Tap each speaker or speaker set you want to include in your multi-room group.

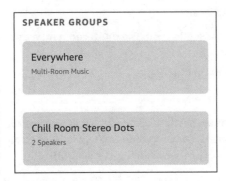

FIGURE 12-8:
Your multi-room music group appears in the Devices screen's Speaker Groups section.

>> "Alexa, play some music on Second Floor."

>> "Alexa, next on Second Floor."

If you want to replay some recent music through the multi-room music group, open the Alexa app, tap Play, tap the music you want to hear in the Recently Played section, and then tap your multi-room music group in the Play On list that appears.

To control the volume on the multi-room music group, include the group name in your voice commands, as shown in these examples:

>> "Alexa, turn down the volume on Second Floor."

>> "Alexa, volume 7 Second Floor."

Fixing multi-room music sync problems

TECHNICAL STUFF

If your multi-room music setup includes one or more Alexa devices that use external Bluetooth speakers, you may find that the playback over the Bluetooth speakers is out of sync with the rest of the Alexa devices. The result is an annoying echo effect. To fix the problem, follow these steps to resync the speaker playback:

1. **Play some music to your multi-room music group.**

   As you see a bit later in Step 7, you adjust the playback of the Bluetooth speaker to get it in sync with the rest of your multi-room devices, so you need to have the music playing to do that.

2. **In the Alexa app, tap Devices.**

3. **In the Speaker Groups section, tap the multi-room music group you're having problems with.**

4. **Tap the Audio Sync Problems? link.**

   The Alexa app displays the Audio Sync screen.

5. **In the Internal Speakers section, decide which device you want to use as a reference for the sync. Leave that device's switch set to on, and tap the switches to off for every other device in the section.**

   You can use multiple devices if you want, but it's usually simpler to work with one device at a time.

6. **In the Connected Speakers section, tap the device you want to compare to the reference speaker you selected in Step 5.**

   If you have just one device with a connected Bluetooth speaker, the Alexa app selects the device for you automatically.

7. **Drag the slider left or right to try to get the playback in sync.**

   The Alexa app adjusts the playback to let you hear the results.

   If you get the playback in sync, skip to Step 9; otherwise, continue with Step 8.

8. **If the playback is still out of sync, repeat Steps 5 through 7 as needed with different combinations of internal and connected speakers.**

9. **Tap Apply.**

Home Sweet Home Theater: Combining Alexa with Fire TV

Another useful Alexa trick is to combine an Alexa-enabled speaker or speaker set with a Fire TV device to create a pretty sweet home theater setup. Here are the steps to follow to make it happen:

1. **In the Alexa app, tap Devices.**

2. **Tap + (add) in the upper-right corner.**

3. **Tap Combine Speakers.**

4. **Tap Home Theater.**

   The Alexa app prompts you to select a Fire TV device.

5. **Tap the Fire TV device you want to use, as shown Figure 12-9; then tap Next.**

   The Alexa app now prompts you to select the speakers you want to connect with Fire TV.

FIGURE 12-9:
Tap the Fire TV device you want to include in your home theater.

6. **Tap each speaker or speaker set; then tap Next.**

   The Alexa app prompts you for a name. The app displays a few predefined names, such as Home Theatre and Entertainment Center.

7. **Tap one of the existing names or use the Customized Name text box and make up your own name; then tap Next.**

The Alexa app prompts you to add your home theater to a group.

8. **Either tap an existing room or group or enter the name of a new one.**

9. **Tap Save.**

The Alexa app sets up your home theater.

In the Devices screen, you now see your home theater in the Speaker Groups section, as shown in Figure 12-10.

FIGURE 12-10:
Your home
theater appears
in the Devices
screen's Speaker
Groups section.

Automating Alexa with Routines

Out of the box, Alexa is a one-command/one-action assistant. That is, when you give Alexa a command, it does just one thing in response to the command — gives you a traffic report, sets a timer, turns on a light, and so on. With scenes (see Chapter 11), rooms, groups, speaker sets, multi-room music, and home theater, you can get Alexa to operate on multiple devices at once, but Alexa is still performing only a single action at a time.

However, there may be times during the day when you ask Alexa to perform a bunch of actions in quick succession. For example, when you wake up, you might ask Alexa to turn on some lights, tell you the weather, play your Flash Briefing, give a traffic report, read your day's events, and then play some music. That's a lot of commands to bark out, especially when you're only half-awake.

Wouldn't it be nice to just say something simple such as, "Alexa, good morning," and then have Alexa automatically run through all those wake-up chores? Yes, it would! Do you need to learn some impenetrable programming language to do it? No, you don't!

Instead, you can create an Alexa *routine*, which is a list of actions that you want Alexa to perform either in response to a single voice command (such as "Alexa, good morning"), at a specified time (such as 7:00 a.m.), or when a device trigger goes off (such as a motion or contact sensor detecting when you enter a room). Alexa comes with a couple of prebuilt routines, but you can create custom routines with not too many taps.

Understanding routine actions

What kinds of actions can you automate into a routine? Amazon is adding new types all the time, but as I write this, the following types of actions can go into a routine:

» **Alexa Devices:** Set the Alexa device's volume.

» **Alexa Says:** Tell Alexa to say something, which can be a preset phrase such as "Good morning" or "Your voice is magnificent" (no, I'm not making that one up) or a custom phrase you enter. You can also get Alexa to sing a song, tell a joke, or relate a story.

» **Audible:** Play from one of your Audible audiobooks.

» **Calendar:** Find out what's on your calendar.

» **Customized:** Type a request that you want Alexa to run.

» **Date and time:** Get the date or time or both.

» **Device settings:** Control some device features, such as the volume.

» **Email:** Hear a summary of your emails.

» **Fire TV:** Control your Fire TV, if you have one.

» **IFTTT:** Run an IFTTT applet (see "Working with IFTTT Applets," later in this chapter).

» **Messaging:** Send or receive a message.

» **Music & Podcasts:** Specify a song, artist, playlist, or podcast.

» **News:** Play your Flash Briefing.

» **Skills:** Open one of your enabled skills.

» **Smart Home:** Specify which smart-home device, group, or scene you want Alexa to turn on or off or adjust.

» **Sounds:** Play a sound effect.

» **Traffic:** Get a traffic report.

» **Wait:** Pause between actions for a specified time.

» **Weather:** Get a weather report.

You can include any number of each type in your routines.

Creating a routine

If you're ready to program Alexa to run a routine, here are the steps you have to follow to make it so:

1. **In the Alexa app, choose More ⇨ Routines.**

You see the Routines screen. The Features tab includes several ready-to-run routines, such as Start My Day and Goodnight, as shown in Figure 12-11. I think you're better off ignoring these existing routines and just creating your own, but you're free to adjust these routines as needed.

2. **Tap + (Create Routine) in the upper-right corner.**

The Alexa app displays the New Routine screen, which has three commands: Enter Routine Name, When This Happens, and Add Action.

3. **Tap Enter Routine Name, type the name you want to use, and then tap Next.**

4. **Tap When This Happens.**

5. **Tap an option to specify how you want to trigger the routine.**

- *Voice:* Enter a voice command to trigger the routine.

- *Schedule:* Select the time when you want the routine to run.

- *Smart Home:* Select the device you want to use to trigger the routine.

- *Alarms:* Run the routine when an alarm is dismissed.

    You might see other options, depending on which Alexa devices you've installed.

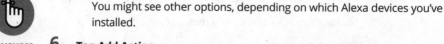

REMEMBER **6.** **Tap Add Action.**

The Alexa app displays the Add New screen, shown in Figure 12-12.

FIGURE 12-11:
The default Routines screen includes two prefab routines.

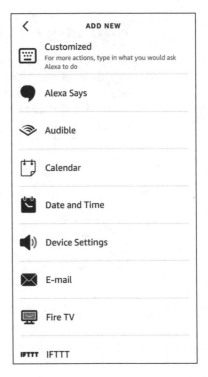

FIGURE 12-12:
Use the Add New screen to add an action to your routine.

7. Tap the type of action you want to add to your routine.

8. Specify the action details, if any, and then tap Add.

9. Repeat Steps 7 and 8 as needed to add all the actions you want your routine to perform.

10. Use the From list to choose which Alexa device you want Alexa to respond from.

If you chose Voice in Step 5, the default is The Device You Speak To, which means whatever Alexa device you say the command to. Otherwise, you can select any Alexa device.

11. Tap Save.

The Alexa app creates your routine.

Running a routine

Unless you created a routine based on a schedule or a device trigger, you run a routine by giving Alexa the voice command you specified. However, you can run any of your routines at any time in the Alexa app also by choosing More ⇨ Routines and then tapping the play icon that appears to the right of the routine, as shown in Figure 12-13.

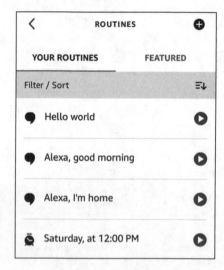

FIGURE 12-13:
You can run a
routine at any
time by tapping
its play icon.

Getting to Know Alexa Guard

With so many people these days investing in smart doorbells, smart outdoor cameras, and smart motion detectors, the phrase *smart home* is becoming synonymous with *safe home*. No surprises there: Your home is your castle and something like an outdoor security camera connected to an Echo Show can let you know if a Trojan Horse is lurking outside your door.

This safety surge might be why the folks at Amazon came up with a service called Alexa Guard, which offers a fistful of features designed to enhance the security and safety of your home.

As this book went to press, Alexa Guard and Alexa Guard Plus were available in only the United States.

REMEMBER

Alexa Guard's features fall into three main categories:

» **Emergency helpline:** An emergency service available 24/7 that you can activate with your voice by saying "Alexa, call for help." Doing so connects you to trained operators who can quickly determine the nature of your emergency and then request the dispatch of the appropriate first responders (ambulance, fire, police, and so on).

» **Smart alerts:** Notifications that you receive on your mobile phone based on sounds picked up by your Alexa-enabled device. For example, you receive a smart alert if Alexa picks up the alarm of either a smoke detector or a carbon monoxide detector. Alexa will also issue a smart alert if it hears glass breaking, footsteps outside your home, or a door closing.

» **Deterring unwanted visitors:** Features that help ward off potential intruders. For example, Alexa Guard can turn your smart lights on and off in a way that makes it look like you're home. If Alexa is connected to an outdoor camera and detects motion outside, Alexa can play the sound of a dog barking or even blast a siren.

These are great features, but the even better news is that many of them are free! That's right: The basic version of Alexa Guard costs precisely nothing. If you want the full set of security and safety features — including the emergency helpline and the barking dog — you need to shell out an extra $5 a month for Alexa Guard Plus. Table 12-1 lists the features of Alexa Guard and Alexa Guard Plus for comparison.

TABLE 12-1   **Alexa Guard versus Alexa Guard Plus**

	Alexa Guard	Alexa Guard Plus
Price	Free	$4.99 per month or $49.00 per year
Emergency Helpline	No	Yes
Smoke detector alarm	Yes	Yes
Carbon monoxide detector alarm	Yes	Yes
Glass breaking sounds	Yes	Yes
Activity sounds	No	Yes
Automatically turn smart lights on and off	Yes	Yes
Play the sound of a dog barking	No	Yes
Play a siren	No	Yes

Working with IFTTT Applets

Alexa is powerful, but sometimes it feels a bit insular because you spend most of your time making Alexa do Alexa-related things. You can help Alexa break out of its shell by connecting it with a service called If This Then That (IFTTT, for short). IFTTT enables connections between disparate apps and devices by using special scripts called *applets*. IFTTT supports hundreds of services, from iOS apps such as Reminders and Calendar to Google services such as Drive and Gmail to devices such as Philips Hue lights and Fitbit.

IFTTT works with Alexa as well, which means you can open up your Alexa voice commands to a wide world of app and device connections:

>> Let Alexa add items to mobile device apps such as Reminders and Calendar.

>> Give Alexa access to extra smart-home device settings.

>> Have Alexa call your phone anywhere in the world.

You can enable dozens of existing applets to make your Alexa experience even more powerful and useful.

To get the ball rolling, surf to `https://ifttt.com` and sign up for an account.

If you're feeling ambitious, you can create your own applets. Go to `https://platform.ifttt.com` and sign up for IFTTT Platform. You don't need any programming know-how. Instead, you create applets using a simple interface where you specify two things:

>> A *trigger*, which is something that initiates the applet, such as an Alexa voice command or action (the "If This" part)

>> An *action*, which is what the applet does when the trigger goes off, such as send an email (the "Then That" part)

Connecting your IFTTT and Amazon accounts

With your IFTTT account activated, you now need to connect your IFTTT account with your Amazon account so that your IFTTT applets can control your Alexa devices. Here are the steps to follow:

1. **Go to** `https://ifttt.com/amazon_alexa`.

2. **Sign in to your IFTTT account.**

3. **Click the big Connect button.**

   You're prompted to sign in to your Amazon account.

4. **Enter your Amazon account credentials, and then click Sign In.**

   You see a description of the access you're granting to IFTTT.

5. **Click Allow.**

   Amazon grants IFTTT permission to access Alexa via your Amazon account.

Enabling an IFTTT applet

With your IFTTT and Amazon accounts joined at the digital hip, you can now activate any Alexa-related applets that look interesting or useful. Here's how:

1. **Go to** `https://ifttt.com/amazon_alexa`.

2. **Sign in to your IFTTT account.**

3. **Locate an applet that you want to enable and click it.**

   IFTTT displays the applet's information page. Figure 12-14, left shows an example page. You get a brief description of the applet; for a bit of extra info, click Learn More.

4. **Click Connect.**

   IFTTT displays the applet's settings. Figure 12-14, right shows the settings that appear for the applet shown in Figure 12-14, left.

5. **Fill in the settings and then click Save.**

   Your IFTTT applet is ready to use.

Checking out some Alexa applets

IFTTT has thousands of applets, so finding interesting or useful examples can be a chore. To help you get started with IFTTT, here are few good Alexa-related applets to try out:

>> **Find My Phone:** If you've left your cellphone in some unknown location around the house, say, "Alexa, trigger call my phone" to have Alexa call your phone so that you can follow the ring tone to locate your lost device. Go to `https://ifttt.com/applets/341784p` to use this applet.

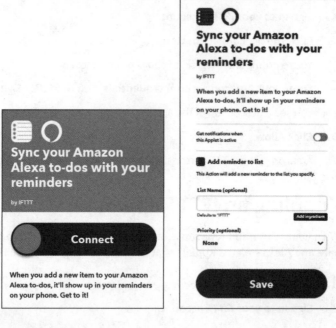

FIGURE 12-14:
The information page for an IFTTT applet (left); most applets have one or more settings you can tweak (right).

>> **Sync your Amazon Alexa to-dos with your reminders.** When you add an item to your Alexa to-do list, this applet also adds the item to your mobile device's reminders list. Go to https://ifttt.com/applets/ieCE52WK to use this applet.

>> **Tell Alexa to email you your shopping list.** Emails your Alexa shopping list to your Gmail account. Go to https://ifttt.com/applets/284243p to use this applet.

>> **Ask Alexa what's on your shopping list, and it will text it to you.** Get your Alexa shopping list items sent to you via an SMS text message. Go to https://ifttt.com/applets/cBkh79yu to use this applet.

>> **Tell Alexa to start the party with a Hue light color show.** Plays a light show that changes colors on a continuous loop using Philips Hue smart lightbulbs. Go to https://ifttt.com/applets/yP4BgwuS to use this applet.

>> **Receive a notification on your phone when your Amazon Alexa timer goes off.** Set an Alexa timer, and when it ends, the IFTTT app on your mobile device displays a notification. Go to https://ifttt.com/applets/303974p to use this applet.

5
The Part of Tens

Chapter **13**

Ten (Times Ten) Ridiculously Fun Alexa Tricks

Your Alexa interactions will mostly have a serious bent as you command your new smart speaker to tell you the weather, play some music, set a timer, or turn on some lights. However, nothing in Alexa's fine print says *all* your conversations have to be sober and practical. Perhaps surprisingly, Alexa has an extensive fun side that you can check out to lighten the mood, entertain your kids (not to mention the kid in you), or relax after a tough day.

In this chapter, you explore Alexa's whimsical nature as you learn how to get Alexa to tell a joke, pose a riddle, sing a song, and tell a story. You'll also delve into the scarily extensive world of Alexa Easter eggs, where you see that Alexa's pop-culture roots run deep.

Ten Ways to Ask Alexa to Tell a Joke

For a virtual personal assistant, Alexa certainly has what seems like a real sense of humor. It's a decidedly corny sense of humor, to be sure. Want proof? Just make any of the following joke requests:

>> "Alexa, make me laugh."

>> "Alexa, say something funny."

>> "Alexa, tell me a blooper."

>> "Alexa, tell me a joke."

>> "Alexa, tell me a prank."

>> "Alexa, tell me a dad joke."

>> "Alexa, tell me a dirty joke."

>> "Alexa, tell me a knock-knock joke."

>> "Alexa, tell me a 'yo mama' joke."

>> "Alexa, tell me a *topic* joke." Alexa will usually come up with a joke related to whatever you substitute for *topic,* including just about any sport (or the word "sport" itself), any science subject (or "science"), and any holiday (or "holiday"). Two other fruitful joke topics are "Star Wars" and "Star Trek."

TIP

Bonus joke request: "Alexa, tell me an Alexa joke."

Ten Alexa Wordplay Tricks

Programmers and nerds of all persuasions love playing with words, so is it any wonder that Alexa — the creation of coders and other Amazonian nerds — is conversant with many different forms of wordplay? Here are ten to have fun with:

>> "Alexa, tell me a haiku."

>> "Alexa, tell me a limerick." (Also: "Alexa, tell me a *topic* limerick.")

>> "Alexa, tell me an oxymoron."

>> "Alexa, tell me a palindrome."

>> "Alexa, tell me some Pig Latin."

>> "Alexa, tell me a puzzle."

>> "Alexa, tell me a pun."

>> "Alexa, tell me a rhyme." (Also, "Alexa, speak in iambic pentameter.")

>> "Alexa, tell me a riddle."

>> "Alexa, tell me a tongue-twister."

Speaking of tongue–twisters, if you want to hear Alexa twist its virtual tongue around a term that's a full 45 letters long, ask, "Alexa, what's the longest word?" The answer, in case you want to follow along, is *pneumonoultramicroscopicsilicovolcanoconiosis.*

TIP

Ten Ways to Get Alexa to Sing

Alexa talks a good game, but did you know that it can also sing? It's true. Sure, Alexa's singing voice won't win any Grammys, but it's pleasant enough, and Alexa's original songs are often hilarious. (My favorite? "It's Raining in the Cloud.") Here are ten ways to get Alexa to serenade you:

>> "Alexa, sing me a song."

>> "Alexa, sing the ABCs."

>> "Alexa, sing 'Auld Lang Syne.'"

>> "Alexa, sing in Auto-Tune."

>> "Alexa, sing a Christmas carol."

>> "Alexa, sing 'Happy Birthday.'"

>> "Alexa, sing a love song."

>> "Alexa, sing the national anthem."

>> "Alexa, beatbox for me." (Also: "Alexa, rap for me," "Alexa, rap for Mom," or "Alexa, rap for Dad.")

>> "Alexa, yodel for me."

You can ask Alexa to sing specific songs, although the results might not be what you expect! For example, try asking Alexa to sing "Itsy Bitsy Spider" or "The Wheels on the Bus."

TIP

Ten Oddball Things to Ask Alexa to Do

The tasks you ask Alexa to do are mostly utilitarian: "Set an alarm," "Play some music," "Be quiet." But all work and no play makes Alexa a dull virtual personal assistant. To give Alexa something fun to do, try any of the following ten requests:

» "Alexa, drumroll."

» "Alexa, flip a coin." (Also: "Alexa, heads or tails?" "Alexa, spin the dreidel," "Alexa, roll the dice," "Alexa, roll *number* dice," "Alexa, roll a *number*-sided die," or "Alexa, roll *number1 number2*-sided dice.")

» "Alexa, high five."

» "Alexa, make me a sandwich."

» "Alexa, pick a random *object*." (For example, you can replace *object* with *animal, actor, card, city, color, food, name, number, occupation,* or *person*.)

» "Alexa, pretend to be a supervillain." (Also: "Alexa, pretend to be a superhero.")

» "Alexa, rock paper scissors." (Also: "Alexa, rock paper scissors lizard Spock.")

» "Alexa, talk like a pirate." (Also: "Alexa, talk like Yoda," "Alexa, talk like a Klingon," "Alexa, moo like a cow," "Alexa, bark like a dog," or "Alexa, make animal noises.")

» "Alexa, tell me something weird." (Also: "Alexa, tell me a thought from the cloud," "Alexa, tell me a fun fact," "Alexa, tell me a fun fact about *topic*," or "Alexa, tell me a life hack.")

» "Alexa, self-destruct."

Ten Funny Questions to Ask Alexa

You can ask Alexa just about anything, and a surprising percentage of the time it will return a legit answer. That's Alexa's artificial intelligence at work, but just because AI is serious technology that doesn't mean all your questions have to be serious, too. Here are ten fun ways to interrogate Alexa:

» "Alexa, can you do an impression?"

» "Alexa, can you sneeze?"

- >> "Alexa, do you believe in love at first sight?"
- >> "Alexa, how much wood could a woodchuck chuck, if a woodchuck could chuck wood?" (Also: "Alexa, how much wood could a woodchuck chuck, if a woodchuck could Chuck Norris?")
- >> "Alexa, how was your day?"
- >> "Alexa, what are you thinking about?"
- >> "Alexa, what is the meaning of life?"
- >> "Alexa, what should I wear today?"
- >> "Alexa, where is Chuck Norris?"
- >> "Alexa, why did the chicken cross the road?"

Ten Ways to Get Alexa to Tell a Story

Alexa comes with a collection of stories and poems, which can be just the thing to hear on a lazy or cold night. Use any of the following voice commands to hear a performance by Alexa, the Storyteller:

- >> "Alexa, tell me a story."
- >> "Alexa, tell me a funny story."
- >> "Alexa, tell me a *topic* story." (For example, you can replace *topic* with "bedtime," "friendship," "love," or "scary.")
- >> "Alexa, tell me your *title* story." (For example, you can replace *title* with "Batter Up," "Measure Twice," "Puppy Love," "Sea Time," or "Tango Night.")
- >> "Alexa, tell me a tale."
- >> "Alexa, what's your favorite story?"
- >> "Alexa, tell me a poem."
- >> "Alexa, tell me a bad poem."
- >> "Alexa, tell me a 'roses are red' poem."
- >> "Alexa, tell me a *topic* poem."

Ten Ways to Get Personal with Alexa

Alexa sounds friendly enough, but how much do you really know about your cloud-based personal assistant? Do you know its birthday or where it lives? I didn't think so. Here are ten questions you can ask to get to know Alexa a little better:

- » "Alexa, are you a robot?" (Also: "Alexa, are you Skynet?")
- » "Alexa, do you have a last name?"
- » "Alexa, do you have pets?"
- » "Alexa, how much do you weigh?"
- » "Alexa, how tall are you?"
- » "Alexa, what are you wearing?"
- » "Alexa, what is your favorite *object*?" (For example, you can replace *object* with "book," "color," "food," "sport," and so on.)
- » "Alexa, when is your birthday?"
- » "Alexa, where did you grow up?"
- » "Alexa, where do you live?"

Ten Movie Easter Eggs

In the software world, an *Easter egg* is a whimsical program feature that's hidden by default and must be discovered. Alexa, you'll be delighted to know, contains hundreds of Easter eggs. You can get a random one using either of these commands:

- » "Alexa, give me an Easter egg."
- » "Alexa, give me a hard-boiled Easter egg."

For a more targeted Easter egg hunt, you can try specific topics such as TV and music, which I discuss in the next sections. Movies are another great source for Alexa Easter eggs. There are dozens, perhaps even hundreds, so consider the following ten to be a mere appetizer:

- >> "Alexa, open the pod bay doors." *(2001: A Space Odyssey)*

- >> "Alexa, who's on first?" (Abbot and Costello)

- >> "Alexa, release the Kraken!" *(Clash of the Titans)*

- >> "Alexa, klattu barada nikto." *(The Day the Earth Stood Still)*

- >> "Alexa, E.T. phone home." *(E.T. the Extra-Terrestrial)*

- >> "Alexa, define supercalifragilisticexpialidocious" *(Mary Poppins)*

- >> "Alexa, are we in the Matrix?" *(The Matrix)*

- >> "Alexa, what is my mission?" *(Mission: Impossible)*

- >> "Alexa, what is the airspeed velocity of an unladen swallow?" *(Monty Python and the Holy Grail)*

- >> "Alexa, what is the Jedi code?" *(Star Wars)*

Ten TV Easter Eggs

Alexa doesn't have a ton of TV Easter eggs, but here are a few to whet your whistle:

- >> "Alexa, what is your cunning plan?" *(Blackadder)*

- >> "Alexa, who shot JR?" *(Dallas)*

- >> "Alexa, don't mention the war." *(Fawlty Towers)*

- >> "Alexa, Winter is coming." *(Game of Thrones)*

- >> "Alexa, who loves orange soda?" *(Kenan & Kel)*

- >> "Alexa, who loves ya baby?" *(Kojak)*

- >> "Alexa, this is a dead parrot." *(Monty Python)*

- >> "Alexa, more cowbell." *(Saturday Night Live)*

- >> "Alexa, what is the Prime Directive?" *(Star Trek)*

- >> "Alexa, where's the beef?" (Wendy's commercial)

Ten Music Easter Eggs

Music is one of Alexa's strong suits, so you won't be surprised to hear that it has dozens of music-related Easter eggs waiting to be discovered. Here are ten to get you started:

>> "Alexa, badger badger badger badger badger." ("The Badger Song")

>> "Alexa, how many roads must a man walk down?" ("Blowin' in the Wind" by Bob Dylan)

>> "Alexa, do you know the way to San Jose?" ("Do You Know the Way to San Jose?" by Dionne Warwick)

>> "Alexa, I shot a man in Reno." ("Folsom Prison Blues" by Johnny Cash)

>> "Alexa, what does the fox say?" ("The Fox" by Ylvis)

>> "Alexa, hello it's me." ("Hello" by Adele)

>> "Alexa, how much is that doggie in the window?" ("How Much Is That Doggie in the Window?" by Patti Page)

>> "Alexa, have you heard that the bird is the word?" ("Surfin' Bird" by The Trashmen)

>> "Alexa, who is the Walrus?" ("The Walrus" by The Beatles)

>> "Alexa, who let the dogs out?" ("Who Let The Dogs Out" by Baha Men)

Chapter **14**

Ten Things That Can Go Wrong (and How to Fix Them)

The good news about Alexa problems — whether the problems are with the Alexa app or an Alexa device — is that they're relatively rare. On the hardware side, although every Echo is a sophisticated device that's really a small computer (not just a fancy speaker), it's far less complex than a full-blown PC, and so much less likely to go south on you. On the software side (and to a lesser extent on the accessories side, meaning external speakers and smart-home gadgetry), skill developers (and accessory manufacturers) have to build their products to work with only a few similar devices made by a single company. This simplifies things, and the result is fewer problems.

Not, however, zero problems. After all, this is still digital technology you're dealing with. Have you ever used a digital device or software that worked flawlessly all the time? I didn't think so. So, yes, sometimes even Alexa behaves strangely. In this chapter, you investigate ten of the most common problems related to Alexa software and hardware and learn how to solve every one of them.

Troubleshooting 101

TIP

Before getting to the specific problems and their solutions, I want to take you through a few basic troubleshooting steps. Many problems, particularly problems related to your Echo device, can be solved by doing the following three things (each of which I explain in more detail in the sections that follow):

>> Restart your Echo device.

>> Update your Echo device's system software.

>> Reset your Echo device to its factory default settings.

REMEMBER

Try restarting your Echo device to see if it solves your problem. If not, move on to updating the software and see if that helps. If there's still no joy, only then should you try resetting your Echo device to its factory default settings.

Restarting your Echo

If your Echo is having trouble connecting to Wi-Fi, pairing with a Bluetooth device, locating smart-home doodads, or doing any of its normal duties, by far the most common solution is to shut the device down and then restart it. By rebooting the device, you reload the system, which is often enough to solve many problems.

TIP

Restarting sounds easy, but you can examine your Echo with your most powerful magnifying glass and you won't find a power button or a restart switch. So, how do you restart your Echo? Take the low-tech route: Unplug it!

You may be tempted to just plug the Echo back in again right away but hold on a second. The Echo has internal electronic components that take some time to completely discharge. To ensure that you get a proper restart, wait at least three seconds before reconnecting your Echo's power supply.

Updating your Echo

Your Echo device uses internal software to perform all sorts of tasks, including listening for the wake word, recording your voice commands, sending your command to Amazon Voice Service (AVS), and handling the result (such as playing an audio file returned by AVS). If your Echo is acting wonky and restarting the device doesn't help, you can often de-wonkify the device by updating the Echo system software. Sometimes installing a fresh version of the software is all you need to make your problem go away. In other cases, updating the software may fix a software bug that was causing your problem.

Happily, all Echo devices update their software automatically. You'll know when a new software update is ready to install because the light ring pulses blue or you hear Alexa say, "An update for this device is available." However, you can take steps to encourage and force an update. See "Your Echo Device Isn't Using the Latest Software," later in this chapter.

Resetting your Echo

If your problem is particularly ornery, restarting or updating the device won't solve it. In that case, you need to take the drastic step of resetting your Echo. I describe this step as *drastic* because it means you have to go through the setup process again, so head down this road only if restarting and updating your device don't solve the problem.

How you perform the reset depends on your device:

>> **Echo or Echo Dot (first generation):** Use the end of a paperclip (or something with a similar tip) to press and hold down the Reset button, which is located underneath the device on the base. Continue holding down the button as the light ring turns orange, then blue, then off. When the light ring comes back on and turns orange, release the Reset button.

>> **Echo or Echo Dot (second generation):** Press and hold down the Microphone Off and Volume Down buttons at the same time. After about 20 seconds, the light ring turns orange and you can release the buttons.

>> **Echo or Echo Dot (third and fourth generation):** Press and hold down the Action button (pointed out way back in Chapter 3). After about 25 seconds, the light ring turns orange and you can release the button.

>> **Echo Flex:** Press and hold down the Action button. After about 25 seconds, the light ring turns orange and you can release the button.

>> **Echo Plus (first generation):** To perform a quick reset, use the end of a paperclip (or something similar) to press and then quickly release the Reset button, which is located underneath the device on the base. This action restores most of the device's settings to their defaults, but it doesn't touch your smart-home device connections.

To perform a factory defaults reset, press and hold down the Reset button for at least eight seconds. This action restores all device settings to their factory default and removes all your smart-home device connections.

>> **Echo Plus (second generation):** To perform a quick reset, press and hold down the Action button. After about 20 seconds, the light ring turns off and you can release the button.

To perform a factory defaults reset, press and hold down the Microphone Off and Volume Down buttons at the same time. After about 20 seconds, the light ring turns off and you can release the buttons.

>> **Echo Spot or Echo Show:** Choose Settings ⇨ Device Options ⇨ Reset to Factory Defaults.

Alexa or Your Echo Device Is Unresponsive

Perhaps the most teeth-gnashingly frustrating problem you can encounter in technology is when a device — particularly one you paid good money for — just stops working. The device appears to be on, but tapping it, shoving it, gesticulating at it, and yelling at it are all ineffective at making the device respond.

If that happens to Alexa or to your Alexa device, try the following troubleshooting steps, in order:

1. **Wait a few minutes.**

   Sometimes devices just freeze up temporarily and then right themselves after a short break.

2. **Check your Wi-Fi network to make sure it's working properly and that your device is connected.**

   See "You Can't Connect to Your Wi-Fi Network," later in this chapter.

3. **Restart the device.**

   See the "Restarting your Echo" section, earlier in this chapter.

4. **Check to see if your device is using the most up-to-date system software.**

   See "Your Echo Device Isn't Using the Latest Software," next.

5. **Reset your device.**

   See the "Resetting your Echo" section, earlier in this chapter.

Your Echo Device Isn't Using the Latest Software

Your Echo device's light ring indicates a pending software update by using a blue, pulsing light (you may also hear Alexa say that an update is available). If you haven't see the blue, pulsing light ring for a long time (or heard Alexa mention anything about an update), you may be concerned that your Echo isn't using the latest software. Follow these steps to check the installed software version against the version of the most recent update:

1. **In the Alexa app, choose Devices ➪ Echo & Alexa.**

2. **Tap the Echo device you want to check.**

3. **At the bottom of the screen, tap About.**

4. **Make note of the Device Software Version, shown in Figure 14-1.**

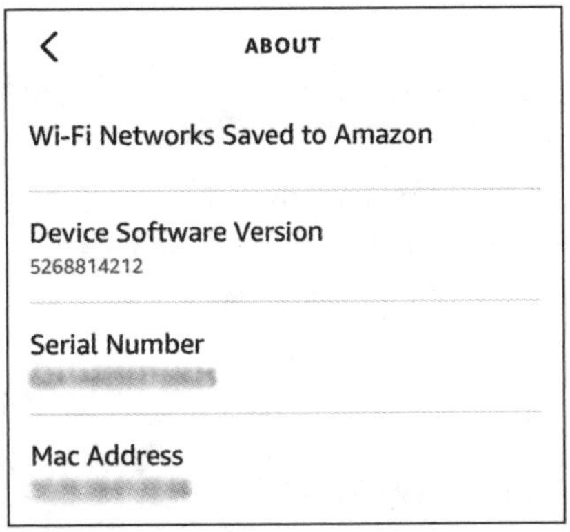

FIGURE 14-1: On your Echo device's About screen, note the device software version.

5. **Go to the Alexa Device Software Updates page at** www.amazon.com/gp/help/customer/display.html?nodeId=201602210.

6. **Find your Echo device type in the list (see Figure 14-2) and make note of latest software version.**

Here's what to do if the two versions are different:

» Make sure your Echo device is connected to your Wi-Fi network. If there's a problem, see "You Can't Connect to Your Wi-Fi Network," later in this chapter.

» Restart your Echo device, as I described earlier in the "Restarting your Echo" section.

Digital Services and Device Support › Echo Device Help › General Echo Support ›

Alexa Device Software Versions

Alexa-enabled devices receive software updates automatically when connected to the Internet. These updates usually improve performance and add new features.

Amazon Echo (1st Generation)
Latest Software Version: 658655620

Amazon Echo (2nd Generation)
Latest Software Version: 665656620

Amazon Echo (3rd Generation)
Latest Software version: 5268811396

Amazon Echo (4th Generation)
Latest Software version: 5268814212

Amazon Smart Oven
Latest Software version: 304093220

Amazon Smart Plug
Latest Software Version: 204000017

Amazon Tap
Latest Software Version: 663643820

FIGURE 14-2:
On the Alexa Device Software Updates page, note the latest software version.

>> Leave your Echo device undisturbed until it gets around to updating itself. Amazon won't update an Alexa device that's in use.

>> If your Echo device has a screen, you can check for an update (and force the update to happen, if one is available) by choosing Settings ⇨ Device Options ⇨ Check for Software Updates.

You Can't Connect to Your Wi-Fi Network

Wireless networking adds a new set of potential snags to your troubleshooting chores because of problems such as interference and device ranges. Here's a list of a few troubleshooting items that you should check to solve any wireless connectivity problems you're having with your Alexa device:

>> **Restart your devices.** Reset your hardware by performing the following tasks, in order:

REMEMBER

Most Wi-Fi devices these days are all-in-one gadgets that combine both a Wi-Fi router and a modem for Internet access. If that's what you have, instead of performing Steps 1–4, you can just turn off the Wi-Fi device, wait a bit, turn the device back on, and then wait for the device to connect to your Internet provider.

1. *Turn off your modem.*

2. *Turn off your Wi-Fi router.*

3. *Turn off your Wi-Fi extender or range booster, if you have one.*

4. *After a few seconds, turn the modem back on and wait until the modem reconnects to the Internet, which may take a few minutes.*

5. *Turn on your Wi-Fi router.*

6. *Turn on your Wi-Fi extender or range booster, if you have one.*

7. *Restart your Alexa device as I describe earlier in the "Restarting your Echo" section.*

>> **Look for interference.** Devices such as baby monitors and cordless phones that use the 2.4 GHz radio frequency (RF) band can play havoc with wireless signals. Try either moving or turning off such devices if they're near your Alexa device or Wi-Fi device.

Keep your Alexa device and Wi-Fi router well away from microwave ovens, which can jam wireless signals.

>> **Check your range.** Your Alexa device may be too far away from the Wi-Fi router. You usually can't get much farther than about 230 feet away from most modern Wi-Fi devices before the signal begins to degrade (that range drops to about 115 feet for older Wi-Fi devices and can go even lower if your home has thick walls). Either move the Alexa device closer to the Wi-Fi router or turn on the router's range booster, if it has one. You could also install a wireless range extender.

>> **Check your password.** Make sure you're using the correct password to access your Wi-Fi network. It's also possible that your Wi-Fi device requires both Wi-Fi Protected Access (WPA) and Wi-Fi Protected Access II (WPA2) for extra security. Alexa can deal with only one of these security types at a time so, if possible, configure your Wi-Fi device to use only WPA2. See your Wi-Fi device documentation to find out how to do this.

>> **Update the wireless router firmware.** The firmware is the internal program that the wireless router uses to perform its various chores. Wireless router manufacturers frequently update their firmware to fix bugs, so you should check whether an updated version of the firmware is available. See your device documentation to find out how this works.

>> **Update and optionally reset your Alexa device.** Make sure your Alexa device is up to date (see "Updating your Echo," earlier in this chapter). If you still can't connect to Wi-Fi, reset your Alexa device (see "Resetting your Echo," earlier).

>> **Reset the Wi-Fi device.** As a last resort, reset the Wi-Fi router to its default factory settings (see the device documentation to find out how to do this). Note that if you do this, you need to set up your network again from scratch.

You're Having Trouble Streaming Media

When you play music, podcasts, audiobooks, or videos through Alexa, the media isn't stored on the Alexa device or even somewhere on your network. Instead, the media file you're hearing or seeing is being sent in real time from Amazon's cloud to your Alexa device, a process known as *streaming*.

Streaming works well most of the time, but problems can arise, particularly the following:

>> The media never starts.

>> The media takes a long time to start.

>> The media plays intermittently.

>> The media stops playing and never resumes.

It's maddening, for sure, but most of the time you can fix the problem. I say *most of the time* because there are a couple of situations where media streaming just doesn't work well:

>> **You have a low Internet connection speed:** Media files are usually quite large, so for these files to play properly you need a reasonably fast Internet connection. Amazon recommends at least a half a megabit per second (0.5 Mbps), but realistically you should probably have a connection that offers at least 5 Mbps download speeds.

>> **You have an intermittent Internet connection:** If you live in an area with spotty Internet service, that now-you-see-it-now-you-don't Internet connection makes streaming media impossible.

What if you have a zippy Internet connection and strong service all the time? First, congratulations! Second, you can try a few things to get media streaming to work better (or at all). Try these troubleshooting ideas in the following order:

>> Restart your Wi-Fi router and Alexa device using the steps I outline in the preceding section.

- » If you have other devices accessing your Wi-Fi network, shut down any devices you're not using.

- » Move your Alexa device closer to your Wi-Fi router. Your Alexa device must be within 230 feet of the router (115 feet for older routers), but the closer the two devices, the stronger the Wi-Fi signal.

- » Make sure your Wi-Fi router isn't situated near devices that can cause interference, such as a microwave oven or a baby monitor.

- » Make sure your Alexa device isn't close to a wall or a metal object.

- » If your Alexa device is sitting on a low shelf or even on the floor, move it to a higher location.

You're Having Trouble Connecting a Bluetooth Device

Alexa devices support a wireless technology called Bluetooth, which enables you to make wireless connections to other Bluetooth-friendly devices, such as headsets, speakers, and smartphones. In theory, connecting Bluetooth devices is supposed to be criminally easy, but in practice that's often not the case. This section provides you with a few common troubleshooting techniques.

You don't see a Bluetooth device

Not surprisingly, you can't make a Bluetooth connection if you can't see the device on the Alexa device's Bluetooth Devices screen (which you display in the Alexa app by choosing Devices ⇨ Echo & Alexa, tapping the Alexa device, and then tapping Bluetooth Devices).

REMEMBER

If you're using an Alexa device with a screen, you display the available Bluetooth gear by swiping down from the top of the screen, tapping Settings, and then tapping Bluetooth.

If you don't see a Bluetooth device in the Alexa app, try the following:

- » Make sure the device is turned on and fully charged.

- » Make sure the device is discoverable. Most Bluetooth devices have a switch you can turn on or a button you can press to make them discoverable.

>> Make sure the Bluetooth device is well within 33 feet of your Alexa device, because that's the maximum range for most Bluetooth devices. (Some so-called *Class 1* Bluetooth devices have a maximum range ten times as long.)

>> If possible, reboot the Bluetooth device. If you can't reboot the device, or the reboot doesn't solve the problem, restart your Alexa device.

>> Check with the Bluetooth device manufacturer to ensure that the device is capable of being paired with Alexa devices. Specifically, you need to find out whether the device supports either of the following Bluetooth profiles:

- Advanced Audio Distribution Profile (A2DP)

- Audio/Video Remote Control Profile (AVRCP)

You can't pair with a Bluetooth device

As a security precaution, many Bluetooth devices need to be paired with another device before the connection is established. You initiate the pairing by tapping the device on the Alexa device's Bluetooth Devices screen (in the Alexa app, choose Devices ⇨ Echo & Alexa, tap the Alexa device, and then tap Bluetooth Devices).

In some cases, the pairing is accomplished by entering a multidigit passkey — sometimes called a PIN — that you must enter. Alexa devices don't support Bluetooth PINs, so if your device requires a PIN, you're out of luck.

Otherwise, you may find that even though the device shows up fine in the Bluetooth Devices screen, you can't pair it with your Alexa device.

First, try the solutions in the preceding section. If you still can't get the pairing to work, tell Alexa to start over by forgetting what it knows about the device:

1. **In the Alexa app, choose Devices ⇨ Echo & Alexa.**

2. **Tap the Alexa device and then tap Bluetooth Devices.**

3. **Tap the downward-pointing arrow to the right of the Bluetooth device name.**

4. **Tap Forget Device.**

   The Alexa app removes the device from the Bluetooth Devices screen.

Press the switch or button that make the device discoverable. When the device reappears on the Bluetooth Devices screen, try pairing with it again.

You Ordered Something Accidentally

The good news with Alexa is that it makes ordering stuff from Amazon as easy as pie. The bad news with Alexa is that it makes ordering stuff from Amazon as easy as pie! That is, when products such as phone soap and moisturizer made from snail mucus are just a voice command away at 2 a.m. on a Sunday, well, it gives the phrase *post-purchase regret* a whole new meaning.

Fortunately, many types of Amazon orders can be canceled if — and this is crucial — you act fast.

WARNING

Some Amazon orders are final sales, including anything you buy from Amazon Music.

If you order something from Amazon.com (or your home-country equivalent), you can cancel it as long as it hasn't gone too far into the processing pipeline. Here are the steps to follow:

1. **Go to** www.amazon.com **(or your local Amazon domain) and sign in to your account.**

2. **Choose Account & Lists (or Your Account) ⇨ Your Orders.**

3. **Locate your order (it should be at the top of the list) and click the order's Cancel Items button.**

   If you see a Request Cancellation button instead, it usually means the item is fulfilled by a vendor other than Amazon, so you need to send your cancellation request to the vendor.

   Amazon displays the items in your order.

4. **Select the check box beside each item you want to cancel.**

5. **Click Cancel Checked Items.**

   Amazon either cancels the order or tells you you're too late.

If it's a Kindle book you don't want, you've got a week to ask for a refund. Here's how:

1. **Surf to** www.amazon.com **(or your local Amazon domain) and sign in to your account.**

2. **Choose Account & Lists (or Your Account) ⇨ Manage Your Content and Devices.**

3. **Click the Actions button to the left of the Kindle book you want to return.**

4. **Click Return for Refund.**

   If you don't see the Return for Refund command, you are too late and are stuck with the e-book.

   Amazon asks you to confirm the return.

5. **Click Return for Refund.**

   Amazon returns the book and processes a refund for you.

Alexa Doesn't Understand Your Commands

Alexa's voice recognition is very good, even downright uncanny sometimes, but that doesn't mean it's perfect. Occasionally, Alexa will not respond at all to a command or will respond with something like "Sorry, I'm not sure about that." Most perplexing of all is when you issue the same command twice in a row, and Alexa works perfectly on the first command but fails on the second!

If this happens to you only every now and then, it's probably just a temporary glitch on Alexa's part, so there's nothing to fix. However, if Alexa consistently fails to understand what you're saying, that's a problem. It makes Alexa not only less useful but also less fun.

Here are a few simple tweaks you can try first:

» **Make sure your Alexa device isn't blocked or covered.** The exquisitely sensitive microphone array in your Alexa device can pick up even faint whispers from several feet away, but it may not hear you at all if the device is behind a wall, under a desk, buried under couch cushions, or tucked away in a closet. Bring your Alexa device out into the open for best results.

» **Move your Alexa device away from any device that has people speaking.** If you place your Alexa device close to a TV, radio, or stereo, when you say your wake word, Alexa may pick up voices from the media device and think they're part of the command, which will almost certainly result in a failure. Keep your Alexa device a reasonable distance from any speaker that's emitting voices.

» **Move your Alexa device away from a wall or other large vertical surface.** Such surfaces can cause reverberations that confuse the Alexa device microphone.

>> **Turn down the volume.** If your Alexa device is playing audio at a loud volume (say, level 9 or 10), the device may not hear you over Led Zeppelin, particularly if you're not close to the device. Turn down the volume by pressing the device's physical Volume Down button and then retry your command.

>> **Check out what Alexa thinks you said.** In the Alexa app, choose More ⇨ Settings ⇨ Alexa Account ⇨ History. Locate the recording where Alexa misheard you and read the transcription of what Alexa thinks you said. If you find Alexa is having trouble with a particular word, ask a friend if you're pronouncing that word correctly. Otherwise, if possible, replace the problem word with a synonym or similar word.

>> **Restart your Alexa device.** I describe how to do this earlier in this chapter in the "Restarting your Echo" section.

If you're still having trouble getting Alexa to understand you, go through Alexa's Learn My Voice feature to train Alexa to recognize you. If you have another Alexa device nearby, turn off its microphone temporarily. Make sure you're in a relatively quiet place, and then give the following command:

"Alexa, learn my voice."

Alexa then takes you through voice training, where it asks you to repeat a series of phrases. Be sure to speak in a natural, conversational tone. Don't speak overly slowly or with exaggerated pronunciation.

If, after all that, Alexa *still* doesn't understand you, try resetting your Alexa device to its factory defaults, as I explain earlier in the "Resetting your Echo" section.

You're Having Trouble Making Calls

If you use Alexa to make voice or video calls, you may find that those calls don't go through at all or go through but suffer from poor quality or frequent disconnects. Bummer. Here are a few things you can try to solve your calling woes:

>> Check your Alexa Contacts list to make sure you're asking Alexa to call the correct name. If you've just edited the person's name in your mobile device's Contacts app, give Alexa a few moments to get the updated info.

>> If you're asking Alexa to call a specific phone number, double-check the number.

WARNING

Alexa won't dial 911 (or any other emergency service number), toll numbers, premium-rate numbers, or *N11* service codes (such as 311 and 411).

» Sign out of your Amazon account. In the Alexa app, choose More ⇨ Settings, and then tap the Sign Out link at the bottom of the Settings screen. Sign back in and try your call again.

» Restart your Wi-Fi router and Alexa device using the steps I outline earlier in the "You Can't Connect to Your Wi-Fi Network" section.

» If you have other devices accessing your Wi-Fi network, shut down any devices you're not using.

» Move your Alexa device closer to your Wi-Fi router. Your Alexa device must be within 230 feet of the router (115 feet for older routers), but the closer the two devices, the stronger the Wi-Fi signal.

» Make sure your Wi-Fi router isn't situated near devices that can cause interference, such as a microwave oven or a baby monitor.

You Changed Your Wi-Fi Network Password

Your Alexa device needs access to your Internet-connected Wi-Fi network to do its thing. When you first set up your Alexa device, getting the device on your network by entering your network password is one of the first chores. That task usually works flawlessly, but what happens if down the road you change your network password? How you tell Alexa about the new password depends on the type of device you have.

If you have an Alexa device with a light ring, you must go through the entire setup procedure all over again:

1. **In the Alexa app, tap the Devices icon.**

2. **Tap Echo & Alexa and then tap the Alexa device you want to connect.**

3. **Tap the Change link that appears to the right of the Wi-Fi Network setting.**

4. **From here, follow the setup instructions that I provide in Chapter 2.**

If you have an Alexa device with a screen, follow these steps:

1. **Swipe down the from the top of the screen and then tap Settings.**

2. **Tap Network.**

   You see a list of available networks, with your network at the top of the list.

3. **Tap the *i*-in-a-circle icon that appears to the right of your network name.**

4. **Tap Forget.**

   The Alexa device disconnects from the network and returns you to the list of available networks.

5. **Tap your network.**

6. **Enter your new network password.**

7. **If prompted, verify your Amazon account.**

A Smart-Home Device Doesn't Work Properly

If you're having trouble with a smart-home device, the good news is that many of these problems typically have a limited set of causes. You may be able to get the device back on its feet by attempting a few tried-and-true remedies:

>> **Check connections, power switches, and so on.** Some of the most common (and most embarrassing) causes of hardware problems are the simple physical things, so make sure that the device is fully plugged in to its power outlet (if applicable), the device is turned on, and any cable connections are secure.

>> **Replace the batteries.** Smart-home devices can really chew through batteries, so if such a device is working intermittently (or not at all), try replacing the batteries to see if that solves the problem.

>> **Turn the device off and then on again.** You power cycle a device by turning it off, waiting a few seconds for its electronic innards to stop humming, and then turning it back on. You'd be amazed how often this simple procedure can get a device back up and running. For a device that doesn't have an on-off switch, try unplugging it from the power outlet, unscrewing it (if it's a light-bulb), or taking out the batteries and putting them back in.

- » **Double-check that the smart-home device is compatible with your Alexa device.** For Alexa units that don't have a built-in hub, you can use only Wi-Fi and Bluetooth devices; for an Echo Plus or second-generation Echo Show (both of which do have hubs), you can use Wi-Fi, Bluetooth, or Zigbee devices.

- » **If applicable, make sure you've installed the smart-home device's app and used the app to set up the device.**

- » **If applicable, use the Alexa app to enable the smart device's skill.** If you've already done that, try disabling the skill and then enabling it again.

- » **Move your Wi-Fi smart-home device closer to your Wi-Fi router.** Your device must be within 230 feet of the router (115 feet for older routers), but the closer the two devices, the stronger the Wi-Fi signal.

- » **Restart your Wi-Fi router and Alexa device using the steps I outline earlier in the "You Can't Connect to Your Wi-Fi Network" section.**

- » **For a Wi-Fi smart-home device, make sure your Wi-Fi router isn't situated near devices that can cause interference, such as a microwave oven or a baby monitor.**

- » **Make sure your Alexa device isn't close to a wall or a metal object.** If your Alexa device is sitting on a low shelf or even on the floor, move it to a higher location.

- » **Reset the device's default settings.** If you can configure a device, perhaps some new setting is causing the problem. If you recently made a change, try returning the setting back to its original value. If that doesn't do the trick, most configurable devices have some kind of restore default settings option that enables you to quickly return them to their factory settings.

- » **Upgrade the device's firmware.** Some devices come with *firmware* — a small program that runs inside the device and controls its internal functions. Check with the manufacturer to see if a new version exists. If it does, download the new version and then see the device's manual to learn how to upgrade the firmware.

Chapter **15**

Ten Ways to Beef Up Security and Privacy

A lexa devices are part of the Internet of Things (IoT), which refers to the collection of gadgets and appliances that have Internet connections. Unfortunately, IoT devices have a well-deserved notoriety for being unsecure. Online miscreants are having a field day compromising so-called smart refrigerators, TVs, and even baby monitors. Does that mean your Alexa device is vulnerable? It's *probably* not, but I'm hedging my security bets here because any system is only as secure as its most vulnerable component. So, even if Alexa itself is fine, you can still run into trouble if your network or your Amazon account are not as secure as they could be. Don't worry, though: In this chapter, I show you how to plug such security holes.

Okay, what about privacy? After all, you've invited a dedicated *listening device* into your home, a device that, moreover, sends recordings of your voice to some cloud and stores them there. Is Alexa *always* listening? Is Alexa recording *everything* you say? As I show in this chapter, Alexa isn't the surveillance nightmare that some privacy advocates say it is, but neither should you take your privacy for granted. There are steps you can (and should) take to protect your personal data, and you learn all about them here.

Make Sure Your Wi-Fi Network Is Locked Up Tight

The first step in securing Alexa is securing the network that Alexa uses to access the Internet: your home Wi-Fi network.

A secure Wi-Fi network is necessary because of a practice called *wardriving*, where a nefarious hacker drives through various neighborhoods with a portable computer or another device set up to look for available wireless networks. If miscreants find an unsecured network they use it for free Internet access (such a person is called a *piggybacker*) or to cause mischief with shared network resources (such a person is called a *jerk*).

The problem is that wireless networks are inherently insecure because the wireless connection that enables you to access Alexa from the kitchen or the living room can also enable an intruder from outside your home to access the network. Fortunately, you can secure your wireless network against these threats with a few tweaks and techniques.

REMEMBER

Most of what follows here requires access to your Wi-Fi router's administration or setup pages. See your router's documentation to learn how to perform these tasks.

>> **Change the router's administrator password.** By far the most important configuration chore for any new (or old, for that matter) Wi-Fi router is to change the default password (and username, if your router requires one). Note that I'm talking here about the administrative password, which is the password you use to log on to the router's setup pages. This password has nothing to do with the password you use to log on to your Internet service provider (ISP) or to your wireless network. Changing the default administrative password is crucial because it's fairly easy for a nearby malicious hacker to access your router's login page and all new routers use common (and therefore well-known) default passwords (such as *password*) and usernames (such as *admin*).

» **Change the Wi-Fi network password.** Make sure your Wi-Fi network is protected by a robust, hard-to-guess password to avoid unauthorized access. See the sidebar "Coming up with a strong password," a bit later in this chapter.

» **Beef up your Wi-Fi router's encryption.** To ensure that no nearby mischief-maker can intercept your network data (using a tool called a *packet sniffer*), you need to encrypt your wireless network. Some older routers either have no encryption turned on or use an outdated (read: insecure) encryption called Wired Equivalent Privacy (WEP). The current gold standard for encryption is Wi-Fi Protected Access 3 (WPA3), so make sure your router uses this security type. If your router offers only WPA2, that's fine, but you might want to consider upgrading your router to a newer model that supports WPA3.

» **Check your network name for identifying info.** Make sure the name of your Wi-Fi network — known as its *service set identifier* (SSID) — doesn't include any text that identifies you (for example, *Joe Flaherty's Network*) or your location (for example, *123 Primrose Lane Wi-Fi*).

» **Update your router's firmware.** The internal program that runs the Wi-Fi router is called its *firmware*. Reputable router manufacturers release regular firmware updates not only to fix problems and provide new features but also to plug security holes. Therefore, it's crucial to always keep your router's firmware up to date.

» **Create a separate Wi-Fi network for smart-home stuff.** If your router supports multiple networks, set up one network for your regular web access and a second network for your smart-home devices. Home-automation gadgets tend not to be very secure, so by shunting them (and Alexa) off to a separate network, you keep your regular network safe.

» **Disable Amazon Sidewalk.** Amazon Sidewalk is a shared Wi-Fi network that's supposed to help your Alexa devices work better, but it also extends your network to your neighbors! Forget that privacy nightmare. In the Alexa app, choose More ⇨ Settings ⇨ Account Settings ⇨ Amazon Sidewalk, and then tap the Disabled switch to on.

WARNING

If you create a second Wi-Fi network for your Alexa and smart-home devices, the Alexa app and any device manufacturer apps will be able to work with their respective devices only if the smartphone (or tablet) running the apps is also connected to that network.

COMING UP WITH A STRONG PASSWORD

As I show in this chapter, making Alexa more secure involves setting passwords for three things: your Wi-Fi network and your Wi-Fi router's administration app, which I talk about in "Make Sure Your Wi-Fi Network Is Locked Up Tight," and your Amazon account, which I discuss later (see "Give your Amazon account a bulletproof password"). However, it's not enough to just use any old password that pops into your head. To ensure the strongest security for your Alexa system, you need to make each password robust enough that it's impossible to guess and impervious to software programs designed to try different password combinations. Such a password is called a *strong password*. Ideally, you should build a password that provides maximum protection while still being easy to remember.

Lots of books will suggest ridiculously abstruse password schemes (I've written some of these books myself), but you really need to know only three things to create strong-like-a-bull passwords:

- **Use passwords that are at least 12 characters long.** Shorter passwords are susceptible to programs that just try every letter combination. You can combine the 26 letters of the alphabet into about 12 million five-letter word combinations, which is no big deal for a fast program. If you use 12-letter passwords — as many experts recommend — the number of combinations goes beyond mind-boggling: 90 quadrillion, or 90,000 trillion!

- **Mix up your character types.** The secret to a strong password is to include characters from the following categories: lowercase letters, uppercase letters, punctuation marks, numbers, and symbols. If you include at least one character from three (or, even better, all five) of these categories, you're well on your way to a strong password.

- **Don't be obvious.** Because forgetting a password is inconvenient, many people use meaningful words or numbers so that their passwords will be easier to remember. Unfortunately, this means that they often use extremely obvious things such as their name, the name of a family member or colleague, their birth dates, or their Social Security numbers. Being this obvious is just asking for trouble. Adding 123 or ! to the end of the password doesn't help much either. Password-cracking programs try those.

Delete Your Saved Wi-Fi Passwords

In Chapter 11, I describe Amazon's Wi-Fi Frustration-Free Setup technology, which makes it easier to set up supported devices by automatically connecting them to your Wi-Fi network. That automatic connection works because Amazon stores your Wi-Fi network password on its servers. Amazon has gone to great lengths to ensure that your saved network password is safe:

>> Amazon's Privacy Policy states that it will not share your Wi-Fi password with a third party without your permission.

>> The password is stored in encrypted form on the server.

>> Devices that ask for network access are first authenticated by Amazon.

>> When needed, the password is sent using an encrypted connection.

These security steps are reassuring, but you may still feel more than a little uneasy having the password to your home network stored in the cloud. And, yes, Amazon authenticates third-party devices that want on your network, but can you really be sure that no rogue device can also breach your network?

To allay these justifiable fears, you can delete your saved Wi-Fi passwords from Amazon and, optionally, turn off Frustration-Free Setup. Here are the steps to follow:

1. **Surf to** www.amazon.com **(or your country's Amazon domain) and sign in to your account.**

2. **Click Accounts & Lists ⇨ Content and Devices.**

   On other Amazon domains, you usually click Accounts & Lists ⇨ Manage Your Content and Devices.

3. **Click the Preferences tab.**

4. **Click Saved Wi-Fi Passwords.**

   The Saved Wi-Fi Passwords settings appear, as shown in Figure 15-1.

5. **To remove your saved network password from Amazon, click Delete.**

   Amazon asks you to confirm the deletion.

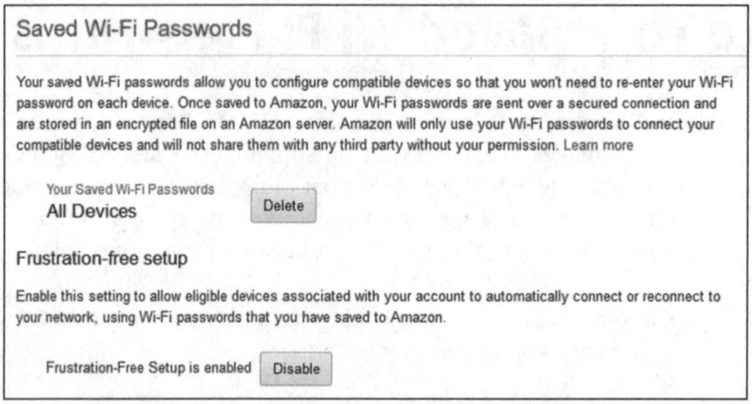

Saved Wi-Fi Passwords

Your saved Wi-Fi passwords allow you to configure compatible devices so that you won't need to re-enter your Wi-Fi password on each device. Once saved to Amazon, your Wi-Fi passwords are sent over a secured connection and are stored in an encrypted file on an Amazon server. Amazon will only use your Wi-Fi passwords to connect your compatible devices and will not share them with any third party without your permission. Learn more

Your Saved Wi-Fi Passwords
All Devices　　　[Delete]

Frustration-free setup

Enable this setting to allow eligible devices associated with your account to automatically connect or reconnect to your network, using Wi-Fi passwords that you have saved to Amazon.

Frustration-Free Setup is enabled　[Disable]

FIGURE 15-1:
Use the Saved Wi-Fi Passwords page to remove your saved passwords from Amazon.

6. **Click Yes, Delete Permanently.**

   Amazon deletes your saved Wi-Fi password.

7. **To prevent devices from using Frustration-Free Setup, click Disable.**

   Amazon asks you to confirm.

   If you proceed with disabling Frustration-Free Setup, you'll have manually add all new smart-home devices to your network.

   WARNING

8. **Click Yes, Disable.**

   Amazon disables Frustration-Free Setup.

Change Alexa's Wake Word

In 2018, an Echo device surreptitiously recorded a conversation between a wife and her husband, and then sent that conversation to one of the husband's work colleagues. Did Alexa have a grudge against the couple, or was it all just a series of unfortunate events? Happily, it seems to have been the latter (Alexa, so far as I know, can hold no grudges). Amazon's explanation for the bizarre occurrence was that the following equally bizarre series occurred:

1. Alexa heard a word in the conversation that sounded like "Alexa," the default wake word, so it began recording the conversation.

2. Alexa subsequently heard a phrase in the conversation that sounded like "send message." Alexa then asked, "To whom?"

3. As the conversation continued, Alexa heard what sounded like the name of a person in the couple's Contacts list. Alexa repeated the contact's name and asked, "Right?"

4. Alexa heard a word — such as "Yes" or "Right" — that it interpreted as confirmation, so it stopped recording and sent the conversation as a message to the contact.

Each of these steps is improbable on its own, but for four of them to happen in a row seems so unlikely that it beggars belief. It *did* happen, though, and the only other explanation is malice aforethought on Alexa's part.

Assuming that Alexa is not a sociopath, what can be done to prevent such an alarming privacy breach? Amazon says it's working on ways to ensure that this series of events doesn't happen again, but in the meantime you can see that it all started with the Alexa device "hearing" the default wake word: "Alexa." So, a good first step is to change the wake word to another word that you're less likely to say.

There are three other possible wake words — Amazon, Echo, and Computer — and you choose one of them by following these steps:

1. **In the Alexa app, tap the Devices icon.**

2. **Tap Echo & Alexa (or All Devices).**

3. **Tap your Alexa device.**

4. **Tap Wake Word.**

   The Alexa app displays the Wake Word screen, as shown in Figure 15-2.

5. **Tap the wake word you prefer to use.**

   The Alexa app warns you that it will take a few minutes to update your device to the new wake word.

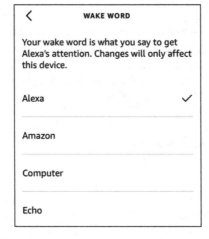

FIGURE 15-2:
Use the Wake Word screen to tap the word you want to use to summon Alexa.

6. **Tap OK.**

   Alexa updates the wake word. Note that until the update is complete, you won't be able to use your Alexa device or the voice component of the Alexa app.

Stop Alexa from Listening (and Watching)

It's unfortunate that Alexa doesn't offer a way to specify a custom wake word. The problem is that the default wake word choices — Alexa, Amazon, Echo, and Computer — aren't that unique, so it's easy to invoke Alexa by accident. Letting you specify an unmistakable wake word — such as, say, "Beelzebub" or "rutabaga" — would go a long way to prevent accidental Alexa wakeups.

Also, you should be aware that your Alexa device really *is* listening to everything you say. The device maintains a constant *audio buffer* — a one-second-long recording — that it monitors for the wake word.

Therefore, to prevent inadvertent wakeups or to prevent that creepy feeling of knowing the device is listening to you *right now,* you turn off the device microphone if you know you won't be using Alexa for a while. You do that by pressing the Microphone Off button, which I point out in Chapter 3.

WARNING

Another good reason to mute your Alexa device when you're not using it is the so-called *dolphin attack* that researchers have demonstrated successfully. Broadcasting audio at a frequency too high for humans to hear (but within the range of dolphin ears, hence the name of the attack), the researchers were able to surreptitiously make Alexa devices dial phone numbers and take photographs.

If you have an Alexa device with a screen, it means you also have an Alexa device with a camera. However, device cameras are notoriously hackable, meaning some creep could gain access to the camera to watch you or take photos. When you mute your device, you also turn off the camera. But when you have the microphone on, it's not a bad idea to be paranoid and turn off the camera if you don't need it. Flick the device's Camera switch to the Off position. If your device has no such switch, you can follow these steps, instead:

1. **Swipe down from the top of the screen.**

2. **Tap Settings.**

3. **Tap Device Options.**

4. **Tap the Camera switch to off.**

Trash Sensitive Alexa Recordings

When you say Alexa's wake word, your Alexa device begins recording everything it hears until your interaction is complete. You may think Amazon just deletes those recordings after Alexa is done with them, but that's not the case. In fact, Amazon keeps your Alexa recordings indefinitely.

Why would Amazon do such a thing? For the simple and innocent reason that Amazon uses your old voice recordings to improve your Alexa experience. Alexa "learns" your voice and your preferences, and that makes your Alexa interactions better every time you use it.

That Amazon has recordings of you asking for the time, the latest Death Cab for Cutie song, or breaking-wind sound effects may seem like no big deal. But as you get more used to having Alexa around, you may expose more private or sensitive information in your conversations, such as medical problems, legal issues, and financial preferences. If having *those* types of recordings stored indefinitely in the cloud somewhere gives you the heebie-jeebies (and it should), you should delete your Alexa recordings.

I'm not saying you have to delete everything. That's probably not a good idea anyway because it would almost certainly mean that your Alexa experience would get much worse for a while as Alexa relearns your voice and preferences. Fortunately, you can be selective about the recordings you delete.

Enabling and using deletion by voice

Perhaps the easiest way to prune your stored recordings is to ask Alexa to delete them for you. Before you can issue commands to remove your recordings, however, you need to activate the deletion by voice feature:

1. **In the Alexa app, choose More ⇨ Settings.**

2. **Tap Alexa Privacy.**

3. **Manage Your Alexa Data.**

4. **Tap the Enable Deletion by Voice switch to on, as shown in Figure 15-3.**

   The Alexa app asks you to confirm.

5. **Tap Confirm.**

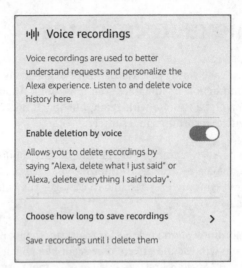

FIGURE 15-3:
To ask Alexa to delete voice recordings, tap the Enable Deletion by Voice switch to on.

With deletion by voice up and running, you can now use the following commands to trash your voice recordings:

>> "Alexa, delete what I just said."

>> "Alexa, delete everything I said today."

>> "Alexa, delete my entire voice history."

WARNING

You might think that saying "Alexa, delete what I just said" would delete only your most recent Alexa command. That would be sensible, but that's not the case here. Instead, Alexa deletes every command you've issued from the last ten minutes.

Configuring Alexa to delete your recordings automatically

If you don't want to be bothered with deleting your recordings manually, you can set up Alexa to do it for you automatically. You can configure Alexa to save your recordings for 18 months or 3 months, after which Alexa trashes them just like that. If you're really paranoid, you can even configure Alexa to not save your recording at all. Here's what you do:

1. **In the Alexa app, choose More ⇨ Settings.**

2. **Tap Alexa Privacy.**

3. **Manage Your Alexa Data.**

4. **Tap Choose How Long to Save Recordings.**

   Alexa opens the Choose How Long to Save Voice Recordings screen, as shown in Figure 15-4.

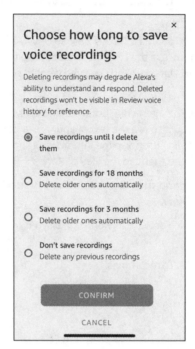

Choose how long to save voice recordings

Deleting recordings may degrade Alexa's ability to understand and respond. Deleted recordings won't be visible in Review voice history for reference.

⦿ Save recordings until I delete them

○ Save recordings for 18 months
Delete older ones automatically

○ Save recordings for 3 months
Delete older ones automatically

○ Don't save recordings
Delete any previous recordings

CONFIRM

CANCEL

FIGURE 15-4:
Use this screen to configure Alexa to delete your voice recordings automatically.

5. **Tap the option you prefer.**

6. **Tap Confirm.**

   Despite just having tapped a button named Confirm, the Alexa app asks you to confirm.

7. **Sigh resignedly and tap Confirm.**

Deleting selected recordings

For the most control over your voice recordings, you can follow these steps to review and remove recordings as needed:

1. **In the Alexa app, choose More ⇨ Settings.**

2. **Tap Alexa Privacy.**

CHAPTER 15 **Ten Ways to Beef Up Security and Privacy** 273

3. **Tap Review Voice History.**

   The Alexa app displays the Review Voice History screen, which by default shows your recordings made today.

4. **Tap Displaying and then use the Filter by Date list to select the recordings you want to review by date.**

   Your choices are Today, Yesterday, Last 7 Days, Last 30 Days, All History, or Custom. If you go with Custom, use the controls that show up to select the day, month, and year for the Start Date and End Date you want to review; then tap Show.

   Each recording shows when the interaction occurred and on which device. Importantly, in most cases you also see the text of your Alexa voice command. If you want to hear the recording itself, tap the interaction's show icon (downward-pointing arrow) and then tap the Play button.

5. **Tap the Filter by Device list and then tap the device you used for the recordings you want to review.**

   If you are not sure of the device or want to work with all your recordings, select All Devices, instead.

   Figure 15-5 shows a typical list of recordings.

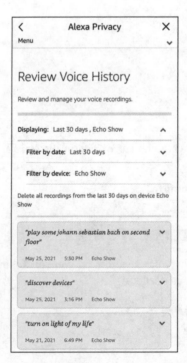

FIGURE 15-5:
The Review Voice History screen shows your Alexa voice recordings from the selected date range.

6. **Choose how you want to delete the voice recordings:**

- To delete individual recordings, tap the recording and then tap Delete Recording or Delete Transcript.

- To delete all recordings in the current date range and device, tap Delete All Recordings from *range* on device *name*, where *range* is the current Filter by Date value from Step 4 and *name* is the current device from Step 5. In this case, Alexa asks you to confirm the deletion.

7. **Tap Confirm.**

The Alexa app deletes your voice recordings.

Configure Alexa Not to Use Your Personal Data

By default, Amazon Voice Services (AVS) uses some of your Alexa-generated data to improve and enhance Alexa. Amazon uses two types of data to make Alexa better:

>> **Voice recordings:** When Amazon is developing a new feature for Alexa, it uses some of its customers' voice recordings to train Alexa on that new feature. The more diverse that collection of recordings is, the more likely it is that the feature will work well for a broad range of users.

>> **Text messages:** When you use a voice command to send a text message through your Alexa device, AVS transcribes that recording into text to ship to your recipient. AVS also uses your messages to analyze the accuracy of its transcriptions.

If you're uncomfortable having your Alexa data used in either or both ways, you can configure Alexa to not include your data when it's improving Alexa. Here are the steps to follow:

1. **In the Alexa app, choose More ⇨ Settings.**

2. **Tap Alexa Privacy.**

3. **Tap Manage Your Alexa Data.**

The Alexa app displays the Alexa Privacy screen, which includes the Help Improve Alexa section, shown in Figure 15-6.

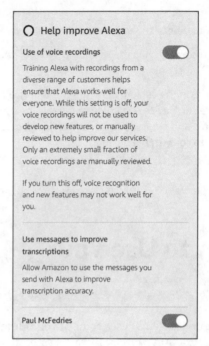

FIGURE 15-6:
Use the switches
on this screen to
control what data
Amazon uses to
improve Alexa.

4. **If you don't want Amazon to use your voice recordings to train Alexa, tap the Use of Voice Recordings switch to off.**

The Alexa app asks you to confirm.

5. **Tap Turn Off.**

6. **If you don't want Amazon to use your text messages, tap the switch that appears beside your name in the Use Messages to Improve Transcriptions section to off.**

Shut Down Alexa's Communications

Alexa's communication features enable you to send messages, make voice or video calls, and drop in (with permission) on other Alexa users or have them drop in on you. It's undoubtedly convenient to be able to perform these tasks hands-free, but they do open up some security and privacy concerns:

>> **Messages are received as an audio file.** That sounds benign, but it may be possible for an attacker to fool Alexa into playing an audio file that contains malicious code.

>> Enabling Drop In on your Alexa device means someone in your Contacts list could initiate a voice conversation or even a video conversation with you without warning.

>> Alexa requires access to your contacts, but if an attacker could somehow manage to infiltrate your Alexa device, your contacts' information would also be compromised.

You can get around these concerns by turning off Alexa's communications features and revoking access to your contacts, as I describe in the next two sections.

Turn communications off

If you don't use (or rarely use) your Alexa device to send or receive messages or drop-ins, you can make your device both more secure and more private by disabling all its communications features. Here's how it's done:

1. In the Alexa app, tap the Devices icon.

2. Tap Echo & Alexa (or All Devices).

3. Tap your Alexa device.

4. Tap Communication.

5. Tap the Communication switch to off, as shown in Figure 15-7.

FIGURE 15-7:
To make your Alexa device more secure and private, tap the Communication switch to off.

<   COMMUNICATION

Enable Communication so you can use Drop In, calling, messaging and announcements on this device.
Learn more

Communication
Disabled for this device

TIP

If you want to only disable the intrusive Drop In feature, leave the Communication switch in the on position and instead tap Drop In and then tap the off option in the Drop In screen.

Revoke access to your contacts

If you've shut down Alexa's communications features permanently, you should also revoke Alexa's access to your contacts because you no longer need them. Revoking access to contacts causes two things to happen:

>> Alexa no longer imports new and changed contacts from your mobile device's Contacts app.

>> The previously imported contacts are deleted from Amazon's servers.

Here's what to do:

1. **In the Alexa app, choose Communicate and then tap the people icon (contacts) in the top-right corner of the Communication screen.**

2. **Tap the manage contacts icon (three dots) in the upper-right corner.**

3. **Tap Import Contacts.**

4. **Tap the Import Contacts switch to off, as shown in Figure 15-8.**

   The Alexa app asks you to confirm.

5. **Tap OK.**

   The Alexa app stops importing your mobile-device contacts, and your previously imported contacts are deleted from Amazon.

FIGURE 15-8:
To remove your contacts from both Alexa and Amazon, tap the Import Contacts switch to off.

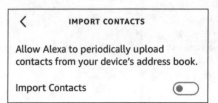

Secure Your Amazon Account

Because everything Alexa does is tied to your Amazon account, your Alexa experience is only as secure as your Amazon account. Therefore, it's vital to ensure that you have your Amazon account locked down. Fortunately, that requires just two things: giving your account a strong password and turning on Amazon's two-step verification feature.

Give your Amazon account a bulletproof password

Your Amazon account's first line of defense is a strong password. First, check back in the "Coming up with a strong password" sidebar. After you have a bulletproof password figured out, follow these steps to change your existing Amazon password:

1. **Surf to** www.amazon.com **(or your country's Amazon domain) and sign in to your account.**

2. **Click Accounts & Lists ⇨ Account (or Your Account).**

3. **Click Login & Security.**

4. **Click the Edit button beside the Password setting.**

   Amazon displays the Change Password page, shown in Figure 15-9.

5. **Type your current password.**

6. **Type your new, strong password in the two text boxes.**

7. **Click Save Changes.**

   Amazon applies the new password to your account.

Your Account › Login & security › Change Password

Change Password

Use the form below to change the password for your Amazon account

Current password:

New password:

Reenter new password:

Save changes

FIGURE 15-9:
Use the Change
Password page to
change your
Amazon account
password.

Turn on Amazon's two-step verification

A password made of steel is a necessary security feature, but, sadly, it's not a sufficient security feature. Malicious users may still worm their way into your account with guile or brute force, so you need a second line of defense. That line is a feature that Amazon calls two-step verification (which is a more comprehensible name than what the rest of the Internet most often uses for the same feature: two-factor authentication). The *two-step* part means that getting access to your Amazon account requires two separate actions:

1. **Sign in using your Amazon account credentials.**

2. **Verify that you're authorized to access the account by entering a code that Amazon sends via text or voice to a phone number you control.**

Here are the steps to follow to enable two-step verification and tell Amazon how you want to receive your verification codes:

1. **Surf to** www.amazon.com **(or your country's Amazon domain) and sign in to your account.**

2. **Click Accounts & Lists ➪ Account (or Your Account).**

3. **Click Login & Security.**

4. **Click the Edit button beside Two-Step Verification (2SV) Settings.**

   Amazon displays the Two-Step Verification (2SV) Settings page.

5. **Click Get Started.**

   Amazon asks how you want to receive your two-step verification codes, as shown in Figure 15-10.

6. **Select the radio button for the method you want to use.**

   - *Phone Number:* Choose this option to receive your one-time passwords (OTPs) via text message on your mobile phone. Enter your mobile phone number, click Continue, get the text code from your phone, type it in the Enter OTP text box, and then click Continue.

   - *Authenticator App:* Choose this option to receive your OTPs using an authenticator app. Use your authenticator app to scan the QR code that appears, type the code in the Enter OTP text box, and then click Verify OTP and Continue.

   Amazon displays some information about using two-step verification on devices that can't display a second screen to enter the verification using two-step verification code.

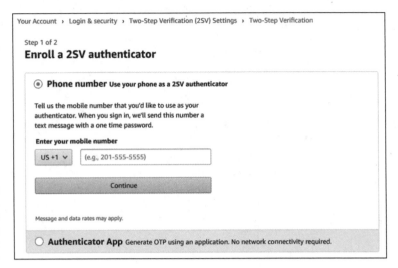

FIGURE 15-10:
Use this page to specify how you want to receive your two-step verification codes.

7. **Click Got It. Turn on Two-Step Verification.**

   Two-step verification is now active on your Amazon account.

Revoke a Skill's Permissions

When you enable a skill, as I describe in Chapter 9, you often have to give the skill permission to access some feature, such as your location, or to perform some action, such as display notifications. Some permissions can get a little annoying (for example, when a skill notifies you too often), but others are more of a problem on the privacy side (for example, when a skill asks for your location). If you no longer want a skill to have whatever permissions you gave it, follow these steps to revoke those permissions:

1. **In the Alexa app, choose More ⇨ Settings.**

2. **Tap Alexa Privacy.**

3. **Tap Manage Skill Permissions.**

   The Alexa app opens the Manage Skill Permissions screen, a version of which is shown in Figure 15-11.

4. **To work with a specific skill, tap Filter by Skill and then select the check box beside the skill.**

5. **Tap the type of permission you want to work with.**

6. **For each skill you want to revoke, tap the skill switch to off.**

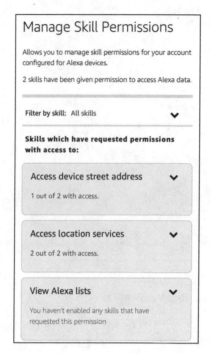

Manage Skill Permissions

Allows you to manage skill permissions for your account configured for Alexa devices.

2 skills have been given permission to access Alexa data.

Filter by skill: All skills ⌄

Skills which have requested permissions with access to:

Access device street address ⌄

1 out of 2 with access.

Access location services ⌄

2 out of 2 with access.

View Alexa lists ⌄

You haven't enabled any skills that have requested this permission

FIGURE 15-11:
Use the Manage Skill Permissions screen to revoke skill permissions.

Deregister Your Alexa Device

If you're selling or giving away an Alexa device, you want to make sure the device is wiped clean of all custom settings and personal info. You can do this by resetting your device, which I describe in Chapter 14.

You also want to ensure that the device is no longer connected to your Amazon account, and that means *deregistering* the device with Amazon. If you have an Echo Show or Echo Spot, you're all set, because resetting the device also deregisters it with Amazon. For other Alexa devices, you have two choices:

» In the Alexa app, choose Devices ⇨ Echo & Alexa, tap the device you're getting rid of, and then tap the Deregister link that appears near the bottom of the Device Settings screen. When the Alexa app asks you to confirm, tap Deregister.

» Surf to www.amazon.com (or your country's Amazon domain), sign in to your account, click Accounts & Lists ⇨ Content and Devices (or Accounts & Lists ⇨ Manage Your Content and Devices), click the Devices tab, click the device category (such as Echo), click the device you're getting rid of, and then click Deregister. When Amazon asks you to confirm, click Deregister.

Index

About the Author

Paul McFedries has been a technical writer for nearly 30 years. He has been talking to — mostly grumbling and yelling at — computers large and small since 1975. He has written more than 100 books that have sold more than four million copies worldwide. Paul invites everyone to drop by his personal website (www.paulmcfedries.com) and to follow him on Twitter (@paulmcf) and Facebook (www.facebook.com/PaulMcFedries/).

Dedication

To Karen and Chase, who make life fun.

Author's Acknowledgments

If we're ever at the same cocktail party and you overhear me saying something like "I wrote a book," I hereby give you permission to wag your finger at me and say "Tsk, tsk." Why the scolding? Because although I did write this book's text and take its screenshots, that represents only a part of what constitutes a "book." The rest of it is brought to you by the dedication and professionalism of Wiley's editorial and production teams, who toiled long and hard to turn my text and images into an actual book.

I offer my sincere gratitude to everyone at Wiley who made this book possible, but I'd like to extend a special "Thanks a bunch!" to the folks I worked with directly: acquisitions editor Elizabeth Stilwell, project editor Susan Pink, and technical reviewer Guy Hart-Davis. I'd also like to give a big shout-out to my agent, Carole Jelen, for helping to make this project possible.

Publisher's Acknowledgments

Acquisitions Editor: Elizabeth Stilwell

Project Editor: Susan Pink

Copy Editor: Susan Pink

Proofreader: Debbye Butler

Technical Editor: Guy Hart-Davis

Production Editor: Mohammed Zafar Ali

Cover Image: © Panthere Noire/Shutterstock

Leverage the power

Dummies is the global leader in the reference category and one of the most trusted and highly regarded brands in the world. No longer just focused on books, customers now have access to the dummies content they need in the format they want. Together we'll craft a solution that engages your customers, stands out from the competition, and helps you meet your goals.

Advertising & Sponsorships

Connect with an engaged audience on a powerful multimedia site, and position your message alongside expert how-to content. Dummies.com is a one-stop shop for free, online information and know-how curated by a team of experts.

- Targeted ads
- Video
- Email Marketing

- Microsites
- Sweepstakes sponsorship

20 MILLION PAGE VIEWS EVERY SINGLE MONTH

15 MILLION UNIQUE VISITORS PER MONTH

43% OF ALL VISITORS ACCESS THE SITE VIA THEIR MOBILE DEVICES

700,000 NEWSLETTER SUBSCRIPTIONS TO THE INBOXES OF

300,000 UNIQUE INDIVIDUALS EVERY WEEK

# of dummies

Custom Publishing

Reach a global audience in any language by creating a solution that will differentiate you from competitors, amplify your message, and encourage customers to make a buying decision.

- Apps
- Books
- eBooks
- Video
- Audio
- Webinars

Brand Licensing & Content

Leverage the strength of the world's most popular reference brand to reach new audiences and channels of distribution.

For more information, visit **dummies.com/biz**

PERSONAL ENRICHMENT

Staying Sharp
9781119187790
USA $26.00
CAN $31.99
UK £19.99

Facebook
9781119179030
USA $21.99
CAN $25.99
UK £16.99

Guitar
9781119293354
USA $24.99
CAN $29.99
UK £17.99

Investing
9781119293347
USA $22.99
CAN $27.99
UK £16.99

Beekeeping
9781119310068
USA $22.99
CAN $27.99
UK £16.99

Digital Photography
9781119235606
USA $24.99
CAN $29.99
UK £17.99

Meditation
9781119251163
USA $24.99
CAN $29.99
UK £17.99

Pregnancy
9781119235491
USA $26.99
CAN $31.99
UK £19.99

Samsung Galaxy S7
9781119279952
USA $24.99
CAN $29.99
UK £17.99

iPhone
9781119283133
USA $24.99
CAN $29.99
UK £17.99

Crocheting
9781119287117
USA $24.99
CAN $29.99
UK £16.99

Nutrition
9781119130246
USA $22.99
CAN $27.99
UK £16.99

PROFESSIONAL DEVELOPMENT

Windows 10
9781119311041
USA $24.99
CAN $29.99
UK £17.99

AutoCAD
9781119255796
USA $39.99
CAN $47.99
UK £27.99

Excel 2016
9781119293439
USA $26.99
CAN $31.99
UK £19.99

QuickBooks 2017
9781119281467
USA $26.99
CAN $31.99
UK £19.99

macOS Sierra
9781119280651
USA $29.99
CAN $35.99
UK £21.99

LinkedIn
9781119251132
USA $24.99
CAN $29.99
UK £17.99

Windows 10
9781119310563
USA $34.00
CAN $41.99
UK £24.99

SharePoint 2016
9781119181705
USA $29.99
CAN $35.99
UK £21.99

Fundamental Analysis
9781119263593
USA $26.99
CAN $31.99
UK £19.99

Networking
9781119257769
USA $29.99
CAN $35.99
UK £21.99

Office 2016
9781119293477
USA $26.99
CAN $31.99
UK £19.99

Office 365
9781119265313
USA $24.99
CAN $29.99
UK £17.99

Salesforce.com
9781119239314
USA $29.99
CAN $35.99
UK £21.99

Coding
9781119293323
USA $29.99
CAN $35.99
UK £21.99